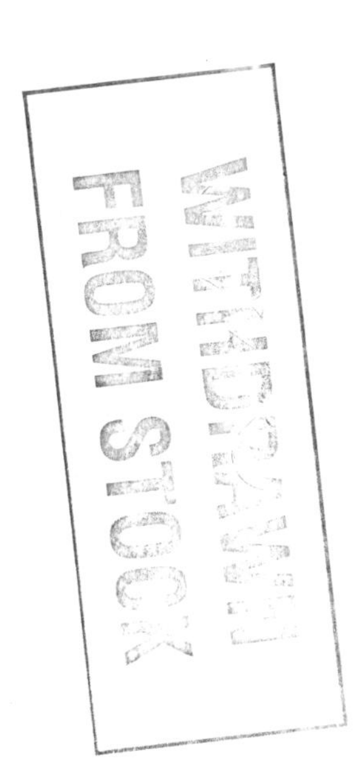

THE CITY

CITY AND SOCIETY

URBAN GEOGRAPHY

CITY AND SOCIETY

An outline for urban geography

R.J. JOHNSTON

LONDON AND NEW YORK

First published in 1980

This edition published in 2007
Routledge
2 Park Square, Milton Park, Abingdon, Oxon, OX14 4RN

Routledge is an imprint of Taylor & Francis Group, an informa business

Transferred to Digital Printing 2007

British Library Cataloguing in Publication Data
A CIP catalogue record for this book
is available from the British Library

City and Society
ISBN10: 0-415-41772-4 (volume)
ISBN10: 0-415-41934-4 (subset)
ISBN10: 0-415-41318-4 (set)

ISBN13: 978-0-415-41772-3 (volume)
ISBN13: 978-0-415-41934-5 (subset)
ISBN13: 978-0-415-41318-3 (set)

Routledge Library Editions: The City

City and Society

An Outline for Urban Geography

R. J. Johnston

London
UNWIN HYMAN
Boston   Sydney   Wellington

Published by the Academic Division of
Unwin Hyman Ltd
15/17 Broadwick Street, London W1V 1FP, UK

Unwin Hyman Inc.,
8 Winchester Place, Winchester, Mass. 01890, USA

Allen & Unwin (Australia) Ltd,
8 Napier Street, North Sydney, NSW 2060, Australia

Allen & Unwin (New Zealand) Ltd in association with the
Port Nicholson Press Ltd,
Compusales Building, 75 Ghuznee Street, Wellington 1, New Zealand

First published in 1980
This updated edition first published by Hutchinson 1984
Second impression 1989

British Library Cataloguing in Publication Data

Johnston, R. J.
City and Society.—Updated ed.
1. Cities and towns
I. Title
910'.091732 GF125

Library of Congress Cataloging in Publication Data

Johnston, R. J. (Ronald John)
City and Society.

Bibliography: p.
Includes index.
1. Cities and towns. 2. Anthropo-geography.
3. Cities and towns—History. I. Title.
GF125.J64 1984 307.7'6 84 – 12821

ISBN 0 04 445429 5

Printed in Great Britain by
Billing & Sons Ltd, Worcester.

Contents

List of Figures 7

Preface 9

Preface to Hutchinson Edition 11

Acknowledgements 12

1. The Concerns of Urban Geography 13

   Spatial organization and spatial science – Criticisms of the spatial organization theme – City and society

2. The Changing Structure of Society 28

   Origins – Mercantilism – Into capitalism – Classes and the capitalist systems – The state and capitalism – Alternatives to industrial and late capitalism

3. The Evolution of Urban Systems: I Pre-Industrial 51

   Urbanization under rank redistribution – Mercantilism and urt systems – System expansion: Mercantilism and colonialism – From model to real world' – Pre-industrial urbanization

4. The Evolution of Urban Systems: II Industrialism 88

   The factory system and location – The factory system and the urban system – Urban growth – Urban systems under industrial capitalism – The onset of late capitalism – Concentration and centralization – The urban system summarized

5. Contemporary Urbanization, Urban Patterns and Urban Problems 123

   Approaches to urban and regional planning – Urbanization and over-urbanization – The problem of urban size – Problem towns and problem regions – Alternative approaches to planning for urban systems

6. Inside the City: The Rationale behind Residential Segregation 153

Classes and industrial and late capitalism – Distancing and residential location – Distancing, territoriality and segregation

7. Housing Classes, Housing Markets and Urban Residential Patterns 185

Contemporary residential segregation – Housing provision and housing classes – Housing market operations and residential segregation – Housing markets and housing choice – The spatial pattern of residential areas – The townscape – Living in the city

8. Non-Residential Land-Use Patterns 233

Initial location patterns – Industries in the city – The industrial pattern, the residential pattern, and commuting – The office sector – Retail land uses – Other non-residential uses

9. Rearranging the City 260

City problems – Environmental problems – Problems of spatial organization – Communities and neighbourhoods – The cycle of poverty and the cycle of affluence – Urban government

Further Reading 287

Index 295

List of Figures

1.1 The rank-size rule 16
1.2 A system of central places 18
3.1 An idealized system of settlements 55
3.2 Two market circuits within one of the cells of 3.1 59
3.3 Two market circuits 60
3.4 Hierarchical shopping behaviour 63
3.5 The hierarchical system with a colony added 68
3.6 The settlement pattern in five colonies 72
3.7 Later development of the pattern in 3.6 72
3.8 Colonialism and urban settlement patterns 75
3.9 Population *v.* rank in the system of 3.1 77
3.10 Population *v.* rank in the system of 3.1, with inter-cell variation in productivity 78
3.11 Four stages in the development of a mercantile urban system 83
4.1 Diagrammatic illustration of the location problem 94
4.2 Diagrammatic representation of the multiplier process 103
4.3 The four components of the multiplier process 107
4.4 Changes in the rank-size ordering of towns in England and Wales, 1801–1911 109
5.1 The relationship between industrialization and urbanization 127
5.2 The costs and benefits of increases in city size 133
6.1 The 'model city' of the South African apartheid policy 178
6.2 The residential pattern of Durban in 1951 and 1970 179
7.1 The Burgess model 216
7.2 The Hoyt sectoral model 217
7.3 The derivation of bid-rent curves 221

7.4 The effect of an increase in commuting costs 222
7.5 The effect of changed preferences on residential location patterns 224
9.1 The cycle of poverty 277
9.2 The cycle of poverty and its links with the cycle of affluence 283

Preface

This book is the product of fifteen years of teaching and research in urban geography. When this period began, urban geographers were very much in the vanguard of human geography, with many of them deeply involved in the pioneering activities of the 'quantitative and theoretical revolution' which was then sweeping through the discipline. Much of their attention focused on economic aspects of urbanization, in particular on the size and spacing of settlements. But the field was expanding rapidly and the social geography of the city soon became a second major focus. Pioneering continued and new areas of interest developed with the study of individual behaviour within the frameworks already described.

It was exciting to be part of this pioneering during the 1960s; nearly every issue of the major journals contained something novel and intriguing for the urban geographer and there was much to be learned from the work of economists, sociologists and, later, psychologists. Research was an immense pleasure, even though a lot of it went wrong, usually because of technical inadequacies on my part. But virtually from the outset I found teaching urban geography much more difficult. The excitement of the research came from the experimentation with techniques and with data sets, all providing bricks for which there was no mortar that would bind them together into an intellectual edifice. In other words there was no real theory around which to hang the empirical work, except for the general belief then being propounded by human geographers that distance was the independent variable with which any pattern could be associated (and thus 'explained'). The biggest difficulty was with the study of urban systems. The theoretical basis was provided by central place theory, which appeared to have little relevance to the real world, certainly not to that part of it in which I was teaching (Australasia). For the social geography of the city, the works of the Chicago school of urban sociologists offered a more promising foundation, but there were still nagging doubts that our data analyses and pattern descriptions lacked any element of deep understanding. The behavioural work was by far the most satisfactory, because the constraints could be assumed and psychological theories and methods applied, cavalierly in many cases, with no great problems: or so it seemed.

The problems were only slowly resolved. The research activity continued unabated, but I was frequently revising my course syllabuses in attempts to provide a viable framework for organizing the research findings that were filling the literature. No acceptable textbooks were produced which provided the needed framework; like my courses, those published comprised a series of case-study essays in urban geography. On the social geography of the city, my own efforts crystallized into *Urban Residential Patterns* (Bell, London, 1971), which was reasonably successful, but my attempt at providing a more general theory in *Spatial Structures* (Methuen, London, 1973) largely failed. And so the search continued, with courses being rewritten frequently and, perhaps as a consequence, research interests straying elsewhere. The major stimulus came from the work in the 1970s which was termed 'radical', and although I have not contributed to the research, its insights have contributed greatly to my organization of courses and helped to provide the framework I had so long been seeking.

Eventually, then, the long haul has borne fruit. The result is this book in which I present, as the subtitle indicates, my attempt at an 'outline theory' which provides an account *for* as well as an account *of* the traditional subject matter of urban geography. Thus the substantial content of this book differs very little from that in other urban geography texts, except that there is not much on behaviour, on individual movements and decisions. (This is because these are very largely conditioned by external constraints, which are the subject of the book.) But there is virtually no technical material. As pointed out already, almost all of the empirical research conducted by urban geographers during the last three decades provides only technically sophisticated descriptions of patterns. I build on those, and support their use in the important process of testing the conjectures advanced in the development of theory. But the emphasis here is on the theory; the materials referred to in the reference section at the end of the book will lead the interested reader to the relevant technical material.

An attempt to write a theoretical outline for a subject as large as urban geography in a book as short as this is bound to simplify, in places probably to over-simplify. Further, in parts it is probably wrong, for as a first essay into such a theory there are many areas where the evidence does not exist and all that I have provided is conjecture. And of course the material can be reinterpreted in the light of competing theories, and the theory I have presented can be revised in the presence of empirical detail which it cannot account for. This is undoubtedly the case with my sketchy treatment of the city in state capitalist societies, but the whole theory is merely a first statement inviting refinement. Indeed the book will have fulfilled its task if it stimulates students of urban geography to consider the theoretical base for urban studies alongside the empirical detail.

Most prefaces end with a statement of acknowledgements for help received, and this is no exception. But first two apologies are required. The first is to the cohorts of students in four universities (Monash, Canterbury, Toronto, Sheffield) who have experienced my various attempts to produce this outline theory: the other is to those many geographers whose ideas I have drawn on (a lot of them would not consider themselves urban geographers, which illustrates the problems introduced by disciplinary name-calling and boundary-drawing) and which are insufficiently recognized in formal references. Perhaps the book will be some consolation to both groups. My thanks are due, as always, to my wife Rita for her reading of the manuscript and, to borrow another author's phrase, 'many other things besides'; to Joan Dunn for her characteristic secretarial efficiency; to Rosemary Duncan, Sheila Ottewell, Peter Morley, John Owen and David Maddison for their cartographic and photographic work on the illustrations; and to Peter Hall and Michael Dover for their editorial advice and encouragement.

Preface to the Hutchinson Edition

In late 1983, the original publishers decided to cease sales of this book, despite the reception since it appeared in late 1980. I am grateful to Peter Hall for his attempts to keep the book in press and to present it to an international market, and to Mark Cohen who agreed to re-publish it. The book has not been revised, but the opportunity has been taken to insert a few additional paragraphs.

April 1984

Acknowledgements

The publishers acknowledge permission to reproduce material in this book from the following sources:

Figure 3.8. From J. E. Vance, Jr, *The Merchant's World* (Prentice-Hall, Englewood Cliffs, New Jersey, 1970). Reprinted by permission of Professor Vance.

Figure 4.4. From B. T. Robson, *Urban Growth: An Approach* (Methuen, London, 1973). Reprinted by permission of Professor Robson.

Figure 5.2. From H. W. Richardson, *The Economics of Urban Size* (D. C. Heath, Lexington, 1974). Reprinted by permission of Professor Richardson.

Figures 6.1, 6.2. From R. J. Davies, 'Changing residential structures in South African cities 1950–1970', in *International Geography 1972/La Geographie Internationale 1972*: Papers submitted to the 22nd International Geographical Congress, Canada (University of Toronto Press, 1972). Reprinted by permission of University of Toronto Press.

1. The Concerns of Urban Geography

Geography is generally accepted by its practitioners as 'that discipline that seeks to describe and interpret the variable character from place to place of the earth as the world of man'; * urban geography attempts such description and interpretation for those parts of the earth's surface classified as urban places. There is no universal definition of urban places. The usual criterion is that an urban area should be a nucleation of settlement in which the great majority of the population is not employed in agriculture. Most statistical agencies set a population size threshold below which such a nucleation is not considered urban. But the majority of urban places are large and the majority of urban dwellers live in big towns and cities. A precise definition is of little relevance except for detailed statistical comparisons, therefore; in this book, the terms 'town' and 'city' are used interchangeably to apply to any non-agricultural settlement containing more than a few hundred inhabitants.

As a recognizable sub-discipline, urban geography is a relatively recent development; the first textbook on the topic in English, for example, was not published until the 1940s. Growth has been rapid in recent decades, however, and urban geography is now a major sub-discipline. The reasons for this late development are not entirely clear, for towns have long been salient features of the landscape. Until the decades after the Second World War much geographical work was concerned with the evolution of landscapes, especially rural landscapes, with a focus on human artifacts rather than on the humans themselves.

Some work on urban areas was done in this genre, emphasizing, for example, the importance of site and situation as influences on urban fortunes. Geographical interest in the definition of regions, areas with common characteristics, led to work on the hinterlands of towns, the areas for which they were the trading nexus, and on morphological regions within towns and cities, as reflected by street patterns, building types and materials, and building uses; interest in the townscape was

* The quotation is from Richard Hartshorne's *Perspective on the Nature of Geography* (Rand McNally, Chicago, 1959), p. 47.

slight, however, in comparison to the volume of work on the rural landscape. Before 1950 very few academic members of the discipline would have called themselves urban geographers, and it was only a change of emphasis in geography as a whole after about 1955 that made the urban place a major focus of academic research.

Spatial organization and spatial science

The conventional methodological and philosophical approaches of geographers were strongly challenged by a new generation of workers in the mid-1950s. Several groups, operating more or less independently in different universities within the United States, led this challenge, and their arguments were rapidly accepted, especially by the many new academic geographers of this period of expansion in higher education.

The stimuli for the challenges were several. Significant among them were the high reputations of the physical sciences and, to a growing extent, some of the social sciences: for the former, both successes in the Second World War and the foundation of the post-war economy on technological advancement ensured their esteem; for the latter, the utility of economic analysis in solving the problems of the 1930s, and of social psychology in armed forces' personnel selection, indicated the potential of scientific social science. Further, with the growing involvement of the state in many sectors of economy and society, there was a great opportunity for academics to participate in planning activities; for geographers, town and regional planning offered such opportunities.

The challenge to the conventional attitudes within academic geography contained a variety of thrusts. The emphasis was on the need to be scientific, to develop laws and theories by using the logical positivist procedures which place great stress on objectivity and neutrality. The discovery of laws allows for scientific prediction, the goal of an increasing number of geographical researchers. Associated with all of this was a proclaimed necessity to use statistical procedures which would ensure the precision and generality of findings. The earlier geography was caricatured as 'mere description', of little utility either in understanding the world as it currently is or in the task of forging a 'brave new world'. A 'theoretical and quantitative revolution' was advanced, therefore, and its proponents had gained control of academic geography within little more than a decade.

The theoretical and quantitative thrusts emphasized philosophy and methodology, especially the latter: the subject matter of the new approach gradually emerged as 'geography as the study of spatial organization'. To investigate not only 'what is done where' but also 'why it is done there' involved accepting that the economic activities carried out in one place are closely linked with those in others, often nearby: the symbiosis between an urban processing plant and the surrounding farms is an obvious example, with the links between the processing plant and a distant market perhaps accounting for that symbiosis. Thus answers to the question 'Why is it done there?', according to the new philosophy, lie not in investigation of each unique case or place but in the search for general laws of location for economic and social activities. Models, theories and laws are common elements in the vocabulary of this 'new geography', and the general descriptive term 'locational analysis' has gained wide currency.

Most societies, and certainly those of North America, Western Europe and Australasia where this 'new geography' developed, focus their activities in and on urban places. The majority of economic functions – in manufacturing, administration and service provision – and of social activities take place in towns and cities, and those which do not, notably agriculture, are articulated through the urban places, on which virtually all routes focus and in which most buying and selling occurs. Thus it is not surprising that the study of urban geography was central to the 'new geography' of the 1950s and 1960s, and this centrality was encouraged by the increased amount of urban planning undertaken.

Urban geographers have developed two main types of research interest during this period of growth in the study of spatial organization, the two types representing analysis at different spatial scales. At the larger scale, they investigate the role of the town within the system of urban places, usually at a national or regional level; at the smaller scale, they analyse the characteristics of different areas within individual towns and cities. These two types are not independent, and methods developed for one have frequently been adopted for the other. Together, they define the current research concerns of most urban geographers.

The study of urban places and urban systems

As indicated above, urban geographers have not been concerned simply to catalogue the activities of the residents of the towns that they study; their aim is to make generalizations about such activities, generalizations which apply to a large proportion of, if not all, towns. A large number of different topics is approached within this broad orientation, but they can be grouped into three characteristic concerns.

1. *The distribution of city sizes.* How many big cities are there in a country or region, and how many small ones? What is the ratio between the population of the largest and the next largest city in a country? Are cities of different sizes growing at the same rates? Questions such as these, involving much collation of statistical data, have frequently been posed as ways of characterizing urban systems and comparing the nature of the urban phenomenon in different areas. Attempts at generalization have focused on two empirical procedures. The first is the so-called rank-size rule, which purports to display a clear relationship between: (i) the size of the largest city in a system (usually a country); (ii) the size of all other cities and towns; and (iii) the position of each place in a rank-ordering of urban populations from largest to smallest. In its simplest form this rule states that the second largest city has a population one half that of the largest, that the third largest has one third, the fourth largest one quarter, and so on; in most cases, these proportions have to be 'weighted' in order to fit the data, but the general relationship holds. (Figure 1.1 illustrates this 'rule'.) The second generalization, the 'law of the primate city', is based on the observation that in many countries the

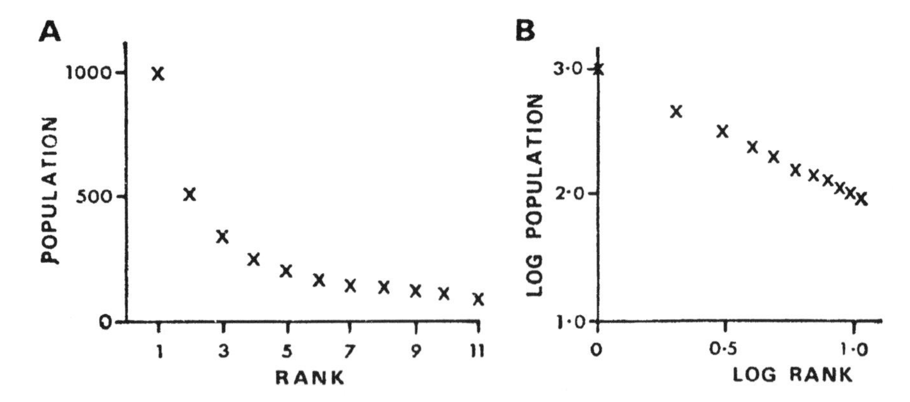

Figure 1.1 The rank-size rule for a system of eleven towns. The largest town has a population of 1,000 and the size of each of the other ten is determined by the formula N/r, where N is the population of the largest town and r is the rank-order position of the town being investigated. (Thus the fourth largest town has a population of 1,000/4=250.) Graph A shows that the distribution of the towns when population is plotted against rank gives a J-shaped curve. If logarithms are taken, for population and for rank, the J is converted into a straight line, as shown in graph B. Different powers for the denominator (e.g., N/r^2 rather than N/r) will produce straight-line relationships of the log curve of varying steepness; the steeper the curve (and thus the greater the power of r) the greater the dominance of the largest town in terms of relative populations.

largest city is at least four or five times bigger than its nearest competitor, and thus dominates the economy and society. And then, of course, some countries display neither a rank-size nor a primate pattern, and so research has sought reasons for the differences, asking whether classifications of countries – large/small; developed/less developed; colony or ex-colony/never a colony; etc. – can be associated with the form of the urban size distribution.

2. *The location of cities of different sizes.* Whereas the work on urban size distributions just described is concerned only with the populations of towns and cities in a given area, in this second category the location of places of different sizes within that area is the major focus. Are places of different sizes apparently randomly distributed across the regional or national space; do they cluster together; or are they arranged in regular sequences?

The major stimulus to work on these questions has been the theories of two German economist/geographers, August Losch and Walter Christaller, both of whom sought to identify regularities in the spatial patterns of urban places. Their arguments were model examples of the positivist method espoused by the new generation of geographers in the 1950s, for their theories were largely deductive, presenting views of how towns should be distributed on the basis of certain assumptions regarding location decisions. These assumptions were that both entrepreneurs (industrialists and shopkeepers) and consumers make rational decisions, and that these decisions involve the minimization of transport costs. (If people have a fixed amount of money to spend on goods and services, then the more they spend on transport – getting to the places where they are sold – the less they can spend on the goods and services themselves. Thus consumers will visit the nearest, and so cheapest, source. To make the profits that they require, entrepreneurs will want people to spend as little as possible on transport, and thus as much as possible in the shops; as a result they will locate as close to the consumers as possible.) Central place theory is built on these assumptions, plus another that there are different types of business requiring markets of different sizes (compare a grocer with a jeweller, for example; the difference reflects the frequency with which consumers purchase from the two).

The predicted urban pattern in Christaller's central place theory has the following salient features: (i) an urban size distribution in which there are groups of towns of the same population, with the larger the size the smaller the number of towns, according to a fixed ratio; (ii) a regular distribution of these towns of different sizes, on a hexagonal network, so that they are as close to the consumers as possible; and (iii) a nesting

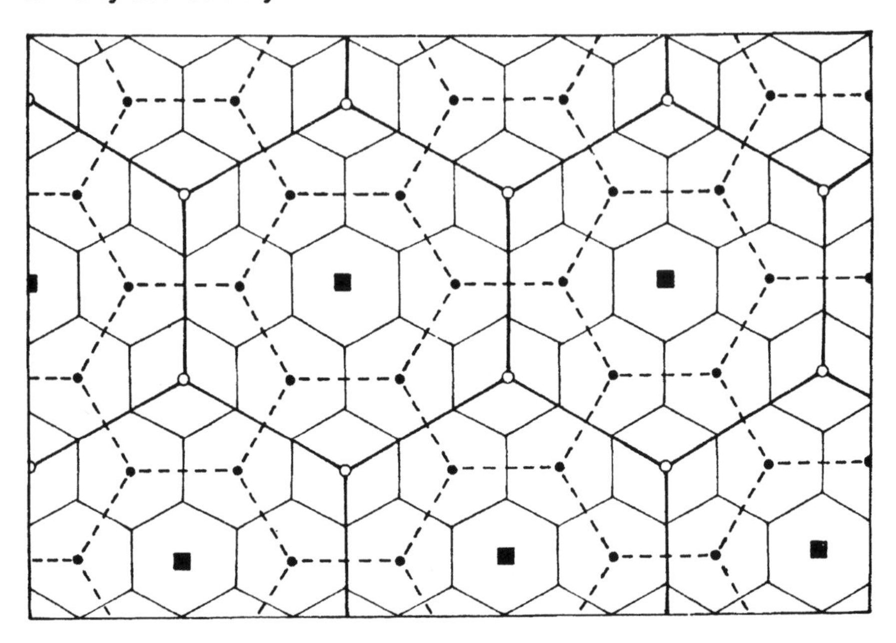

Figure 1.2 A system of central places according to Christaller's model. Three levels are shown in his hierarchy. At the highest level are the towns shown by squares; these serve the hexagonal market areas defined by the heavy lines. At the intermediate levels are the towns shown by open circles, which serve the market areas defined by dashed lines At the lowest level are the towns shown by dots, serving the hinterlands indicated by faint, solid lines. (Note that this is only one of three idealized patterns deduced by Christaller.)

arrangement of the hinterlands of different towns so that, for example, a town in size group A might contain six towns of size group B within its hinterland, and in turn each of these towns of group B will have six towns of size C in its hinterland. The result is a hierarchical spatial system of nesting hinterlands on a hexagonal frame (Figure 1.2), and of links between towns articulated through this hierarchy. Losch's predicted structure is more complex, because of different assumptions regarding types of business, but it has the same major feature, a hexagonal network; its urban size distribution is more akin to a rank-size rule than to a stepped hierarchy, however.

Many geographers were attracted by these theories, especially Christaller's, and sought to establish whether urban patterns can be accounted for by comparison with the hexagonal model. Further, the general thesis that economic activities are located so as to minimize transport costs provided the underlying basis for most work in human geography, and theories of least-cost location were developed to provide the datum against which patterns in the 'real world' could be compared.

3. *Urban functions and urban classifications.* Central place theory, as developed by Christaller, assumes that the dominant function of each urban place is the articulation of trade, both local and long-distance. As developed by Losch, the town is also the location of manufacturing industries, which serve local hinterlands. Both theories present only a partial view of the economic functions of towns, therefore, for there are many functions apparently tied to certain places only, in particular because of the proximity of certain resources. Location of these functions may be determined by the same considerations regarding transport-cost minimization, but the resulting distributions will not accord with the uniform sequences of the hexagonal net.

Economic geographers have been particularly concerned with analyses of the locations of individual industries, whereas urban geographers focus on identifying what functions are performed in what places, usually basing their investigations on employment data. The result has been a series of city classifications, identifying, for example, trading centres, administration towns, transport centres and manufacturing towns, with the last perhaps subdivided according to the types of goods manufactured. In a few cases the different economic functions have been associated in statistical comparisons with various social and demographic characteristics of the urban populations; in others the movement of materials and finished goods has been analysed to indicate the articulation of the interdependent urban places in a system wherein every place is both supplier and customer for every other.

Three major features of all of these types of work can be identified from the urban geographical literature. First, it is strongly descriptive in emphasis: most studies have been presentations of empirical detail, within the general context of location theories emphasizing transport-cost minimization. Second, there has been considerable attention to the sophistication of statistical procedures used in the descriptions, and in the comparisons of observed urban patterns with those predicted by the accepted theory. Finally, the theoretical background has been drawn almost in its entirety from neo-classical economics, emphasizing the primacy of the individual as the decision-maker and the rationality of decision-taking. The models of urban patterns assume both complete information and perfect ability to use that information on behalf of decision-makers, so that they portray, as a datum, what the world's spatial organization should look like in certain idealized circumstances. Such a model is taken to be realistic as a first approximation. Deviations from its predictions are thus the result of irrational behaviour; in the long term decisions made irrationally should prove unsuccessful (a shop in the wrong place, for example), so that the pattern of survivors should conform with the predictions.

Some of this empirical work, and the theory on which it is based, has been used as input for planning future urban systems. Geographers have in particular been called on to provide expertise in data handling, but their models have been used as the goals for planning exercises.

The internal structure of cities

Whereas studies of urban systems treat each place as a point on the economic and social landscape, at the second spatial scale the town is disaggregated and its component parts treated as separate items for analysis. Again, three main types of work can be identified.

1. *The location and characteristics of non-residential land uses.* Two land uses have received most attention: retailing and manufacturing. Central place theory has provided the framework for analysis of the former; the model against which the pattern of shopping centres has been evaluated includes a distribution of centres of different sizes, on a hexagonal network, and organized in a series of nesting hinterlands. In addition to analyses comparing the 'real world' with this model, it has also been adapted to predict the locations of different business types within individual shopping centres. The location of manufacturing has been investigated using the basic premise of least-cost location and, in recent years, the locations of the growing number of offices in cities have been analysed in similar fashion. The result is a classification of areas dominated by different types of either industry or offices.

2. *The characteristics of residential areas.* This is the main subject area studied by urban geographers working at this scale, and embraces a variety of concerns. Considerable interest has been shown, for example, in the general pattern of population distribution. Densities have been found to decline with increased distance from the city centre, and this has been accounted for by the transport-cost minimization thesis, which in turn 'explains' the parallel decline in land values away from the city centre (people are assumed to be prepared to pay more to live close to the centre, so that they will have to pay less to travel to work, to shops and to other destinations).

A major observation of patterns within the city is that different population groups tend to live in separate areas. Various socio-economic status groups, for example, representing the division of labour in an urban society and the different rewards associated with that division, live apart from each other; immigrant and other minority groups frequently occupy highly segregated residential areas; and there is a tendency for the separation of age groups. To demonstrate the various facets of this residential differentiation, the methods of social area analysis and its more sophisticated statistical development, factorial ecology, have been applied to large data sets, thereby providing descriptions of the nature and extent of the separation. Explanation for the observed patterns accepts the division of society into its various groups, and suggests that because members of each group only want to live among, and interact with, people like themselves, thus separation occurs; it is proposed that the highest-status groups, with most income, have first choice of areas.

As well as describing residential differentiation, urban geographers have also analysed where within cities the various groups tend to be located. Their models are taken from American work of the 1920s and 1930s which suggests both a zonal and a sectoral pattern to the differentiation, and, again, sophisticated statistical procedures have been applied in the comparison of particular places against these ideals. Further, the models accept that the characteristics of areas change over time, and that people migrate between them; analyses of these changes, and especially of the migrations, are firmly set in the transport-cost minimization mould.

3. *Traffic flows between areas*. Interest in this topic developed out of the planning requirements of the 1960s, when rapid expansion of automobile ownership in most urban societies produced problems of traffic management. The initial focus was on the traffic-generating capacities of different land uses and on patterns of movement which generally conformed to predictions based on transport-cost minimization hypotheses (i.e., most moves were over short distances). Technical sophistication allowed development of models which both described current travel patterns (notably journeys to work and to shops) and predicted future flows. The latter have had important planning applications, and have been employed to predict the best locations of new residences, workplaces, shops and roads, within the context of cost minimization.

As with studies at the larger spatial scale, work on patterns within cities has been both strongly descriptive and highly technical. Furthermore, much of the theoretical background has been taken from neoclassical economics and assumes a certain type of decision-making: in the spatial context this is translated into decisions aimed at transport-cost minimization, for both residential and non-residential land users. In addition, work on residential patterns has accepted the functionalist theories of sociologists which divide society into a series of competing groups, within which individuals accept their position but strive, by the accepted means, to improve themselves.

Most of the work in the spatial organizational theme within urban geography has been partial in its intent, looking at a particular aspect of the urban phenomenon only, in isolation from all others. (In part this reflects the academic division of labour within geography: the flow of goods between places, for example, is commonly studied by transport geographers.) Attempts have been made to link the various parts into a coherent whole. The most widely accepted framework presents the urban pattern as a system of linked, interdependent parts, analogous to a machine such as the internal combustion engine in which no one part can operate without both all of the others and a supply of energy. Within each place the pattern of land use is similarly phrased as a system of interdependent parts, giving an overall view of an urban place as a system nested within a system of urban places. For more than a decade the main task of urban geographers was describing the operation of such systems and their spatial manifestations.

Description requires a theory to provide its language, and urban geographers drew much of their theoretical inspiration from a certain type of economics, with which some sociological theory was merged when relevant. This theory assumes a society comprising a large number of independent units (firms, households and individuals) whose positions within the relevant economy and society are taken for granted. Decision-making by these units is aimed at the minimization of transport costs. Firms locate their business premises so as to keep to a minimum the costs of bringing in materials and of sending their products to market; households locate their homes so as to reduce, as far as possible, the costs of travelling to work, to shops, to places of entertainment and to the homes of those (friends and kin) they visit frequently.

Criticisms of the spatial organization theme

From the mid-1960s such theory has been subjected to an increasing volume of criticism, largely on the basis of its assumptions, which it is claimed are unreal. The approach to urban geography, like the approach in most other aspects of the 'new' geography, has been criticized in two major ways, although the concerns, the topics studied by urban geographers, remain very much the same.

The behaviourist critique

The focus of this set of criticisms is the pair of assumptions that people wish to maximize their utility and that they do this on the basis of complete information and perfect ability to use it. Most people, it is claimed, aim only at a satisfactory utility level; once that has been achieved they are not prepared to invest the extra effort involved in obtaining a further increase, since the extra benefits would not warrant the costs. (A useful analogy here is the chess-player who, having identified a sequence of moves which will give checkmate in seven, will not then spend a great amount of time investigating whether the same end can be achieved in six moves.) Furthermore, in their search for locations (to live at, to shop at, or as the site for a factory, for example) information is almost always incomplete – people cannot possibly discover every possible potential site and evaluate its characteristics. Nor, with the information that they have and their reasoning abilities, can people be expected to make perfect decisions: they cannot be expected, for example, to predict how their competitors will behave. (And every decision-maker cannot outguess all competitors!) And finally, for a variety of reasons – some peculiar to the individual, some to the cultural group within which he or she has been socialized – aspirations vary, with obvious consequences for locational decision-making.

For these reasons, a number of geographers have argued that to understand spatial organization, and spatial behaviour within the pattern of that organization, attention must be paid to the ways in which people make location decisions. The findings of such investigations can then be used as the axioms on which better location theories are built. Thus a behaviourist (or positive) approach has been advanced, and its research in many cases involves the collection of data from relevant decision-makers by survey procedures, in some cases guided by the theories of learning and of search developed by psychologists for other contexts. The aim was not the destruction of the theories of the spatial organization school, however, but rather their improvement, and so the behaviourist work was structured within the frameworks of, for example, central place theory and social area analysis and used their sophisticated technical arsenal. The unit of analysis was the individual, in attempts to discover, for example, why households migrate within cities and how they select their destinations. In a sense, the research has inquired to what extent the transport-cost minimization assumption is valid.

Structuralist critiques

This second set of criticisms of the spatial organization school is more fundamental, attacking both the axioms of its theories and the use of its results in planning activities. On the first point, both neo-classical economics and functional sociology are attacked for their ahistorical, static approach. The former assumes the existence of independent, competing firms which make locational decisions; the latter assumes the existence of classes. But the nature of firms is continually changing, as the individual units compete for markets and for profits and as the wider economic environment influences their potential for success. Similarly, neither the social classes in existence at any one time nor their membership is immutable; classes evolve as part of the economic, social and political organization of society. The spatial organization approach can map the patterns which reflect the current operation of inter-firm competition and inter-class conflict, but its theories based on cost minimization cannot account either for the evolution of those patterns or for the ways that they will change in the future, since they offer no explanation for why transport-cost minimization is important.

The behaviourist approach is similarly criticized. Its philosophy is reductionist, focusing on the individual as an independent decision-making unit. But individuals operate within constraints imposed by the economic and social environments as well as by the physical and spatial. Without studying those constraints, behaviourist work emphasizes choice when there may be very little of that available. It can account for why certain decisions were taken, out of the alternatives considered, but cannot account for the alternatives. Locational decision-making, like locational patterns, can only be completely accounted for by investigating the mechanisms which constrain the boundaries of individual decision-making, and perhaps for some people ensure that there is no choice at all.

Regarding the use of the results of spatial organization studies in planning urban futures, it is claimed that this cannot produce any substantial change. Because the empirical investigations, and the theories on which they are based, merely describe current patterns and relationships within urban systems, they can only predict future patterns and relationships within the same constraints. The plans affect only the spatial outcomes of economic and social processes, not those processes themselves, and so are conservative in their effect. Detailed spatial forms may be affected but, for example, urban renewal will not change the relative economic and social positions of the inner city poor, only the material environments of their poverty. If the aim of planning is an improved quality of life for all, then it cannot be built on static, descriptive investigations, however successful they might be within their own terms of reference. Successful planning requires an understanding of the constraints to behaviour, of the operation of the fundamental social and economic processes which produce the generality of spatial patterns.

This series of criticisms has engendered a debate within the whole of geography between, on the one hand, the spatial organization school and their behaviourist colleagues and, on the other, the structuralists who argue that study of spatial systems must be set within the context of the encompassing milieux. According to the latter group, the spatial organization approach implicitly accepts the current structure of society and its distribution of economic and political power. It describes the outcomes of the exercise of that power, but does not explain how power relationships circumscribe the framework for individual action. The results of spatial organizational studies pose questions about the manipulation of space by those with power in society, questions which cannot be answered with reference to space, and to transport costs, alone. Any planning based on the descriptions produced must ensure continuation of the power structure, and so is ideologically conservative. Creative change can be achieved only by understanding how society operates to structure space within the urban system, and then producing programmes which will affect the operations, not just the outcomes.

City and society

The concerns of urban geography, then, are the organization of space within an urban context: what economic, social and political functions are conducted in which towns, and why; what trends are being operated to alter this distribution of functions; why different land uses are located in different areas within cities; and how and why these location patterns are changing. To date, much of the work by urban geographers has been concerned with sophisticated (at least in a technical sense) description of patterns and of trends, but there has been little work that offers satisfactory explanations for the urban phenomenon and its many characteristics. The major lack is of a general theory which explains how and why towns play a major role in contemporary society and how and why functions and activities are distributed between and within different urban areas. The present book attempts an outline for such a theory.

Much of the structuralist critique of the types of urban geography practised in recent years, and still practised by most professed urban geographers, is accepted here. The patterns as described are not challenged; they are the basis for the theory, having asked the questions for which an answer is proposed. The basic theme followed here is that the city is a social artifact whose form can only be understood in the context of an understanding of the relevant society. The theory proposed is thus essentially a backward-looking one, seeking to explain the present through the past and treating the current urban phenomenon as the outcome of an evolutionary process. There is no discussion of the future, of what the system should look like.

A general urban theory, when complete, will be a mammoth piece of work and it must be stressed that the present book provides only an outline of the form that theory should take. Of necessity, therefore, what is presented here is highly generalized and lacks detail, although examples are given, most of them from the English-speaking world. Much of the detail is place-specific for, as the structuralist case makes clear, each process allows for local variability; it sets the constraints, or degrees of freedom, within which action can take place according to local interpretations (as, for example, with the amount of public housing in different societies and the varying attitudes to urban sprawl). Thus the theory outlined here indicates how the processes work: it sets the general framework within which particular patterns emerge according to local circumstances, some of which are presented as illustrations. The constraints are reasonably tight, however, otherwise the theory would have little value: local variations influence the detail but not the specific form of the urban phenomenon.

Although an urban system reflects the economy and society which creates and maintains it, it is not entirely passive and lacking in its own existence and influence. A modern urban system, comprising buildings, transport networks and an infrastructure of utilities, is a major investment which cannot easily be replaced. To some extent, therefore, its form influences the evolution of economy and society, usually only slightly but in some cases acting as a major determinant of future forms. The current urban system is thus a palimpsest, a record of societal evolution and the spatial matrix within which it has occurred.

The argument presented here, therefore, is that the urban system both reflects and interacts with the economic, social and political structure of the society which occupies it. Of these three aspects of society, the economic is the dominant one, and both social relations and the distribution of political power are contingent upon the economic mode of organization. Over much of the earth's surface at the present time that economic mode of organization is late capitalism, which has evolved out of a sequence of earlier forms, of which the preceding two modes were mercantilism and industrial capitalism. In the rest of the world alternative modes of organization have been introduced, on a base which was mercantilist. Thus an outline of urban geography must be grounded in an outline of the development of late capitalism, of its social structures and power distributions, and of the other forms which have been superimposed on its predecessors. Such an outline is provided in the next chapter.

Summary

This book is concerned with the traditional topics studied by urban geographers: the sizes, locations and functions of towns, the characteristics of their constituent areas and the location decisions which create, maintain and alter the form of urban systems. Its aim is to provide a general theory which will explain those patterns and processes, and so the emphasis is on the broad outline, on the causes and not the consequences, the common rather than the unique. Thus the sophisticated technical descriptions produced by urban geographers in recent decades are shunned, not because they are wrong but because they are too detailed; they are the material which confirm the outline presented here, but which must be referred to separately.

Following the outline of the evolving structure of economy and society in Chapter 2, the remainder of the book is in two halves, representing the two scales at which urban geographers work. In the first (Chapters 3–5) the urban system is analysed; in the second, the internal structure of the city is the focus of attention (Chapters 6–9). In all of these, there are only a few references to particular examples. The book is very much a hypothesis, therefore, a theory which is internally consistent and apparently relevant to the 'real world' but which awaits detailed testing that will both confirm the general outline and fill in the specific detail. It is offered in that spirit, as an outline of the theory which urban geography so badly needs.

2. The Changing Structure of Society

This chapter provides the context for presenting a theoretical outline of urban geography in the remainder of the book. The presentation of such a theory in only seven chapters is a considerable generalization, so a discussion of the changing structure of society in a single introductory chapter must be even more generalized. Thus what follows is an extremely sketchy outline of the general trends that have resulted in contemporary world societies.

Origins

The first societies were small and occupied limited areas at any one time (many migrated from one area to another, seasonally, annually or after a few years, following the depletion of available resources). All members lived at very low subsistence levels. As far as can be inferred, social organization was highly egalitarian. All active individuals were involved continuously in the tasks of producing food and other basic materials. Decisions were taken democratically, either in meetings of the whole society or by the consensus allocation of particular forms of decision-making to specified individuals.

With such societies there was little exchange of goods; what occurred took place between equals. There were, of course, transfers to those unable to provide the means of subsistence for themselves – the young and the old. The division of labour was a social one only, therefore, except in those cases where there was some delimitation of tasks according to sex. Any exchange was based on reciprocal principles; there was no medium of exchange, and no group within society exerted economic and social power over others.

Almost all of these egalitarian, subsistence societies continued at very low levels of existence. Everything produced was consumed almost immediately, and a fall in production was certain to bring problems of human survival. They were very largely controlled by their environments, and lacked the means to develop techniques which might allow environmental control. (Fire was the first, although this was rarely controllable.)

Because of this dependence on external circumstances, most societies collapsed, either being entirely eliminated or, more likely, being decimated, to revive to the previous low subsistence levels when environmental conditions improved. A few, however, through a combination of chance and organization, were able to expand their volume of production, to store the surplus for later use, and thereby to ensure a substantial level of subsistence for all and perhaps sufficient for above-average existence for at least a few. In such situations egalitarianism almost certainly disappeared, being replaced by some rank-ordering of society's members which was the basis for redistributive mechanisms.

Each producing unit in such redistributive societies was required to supply a certain volume of food and/or other raw materials to specified members. This supply may have been a predetermined absolute volume, a proportion of the total production or a certain amount of work performed on the land of those to whom the tribute was due. To ensure the provision of either goods or labour, the recipients needed some power over the suppliers, to ensure, through coercion if necessary, that rank was respected and rewarded. Such power was usually based on either religious precepts, with a priesthood being able to exert spiritual influence, or on military might, in which an overlord, almost certainly hereditary, coerced goods and labour from subjects. In the latter case landownership in the society was probably vested in the overlord, who allowed his subjects to cultivate part of his territory for their own subsistence, in return for their labour on his portion. Both military overlord and priesthood were present in many societies, occupying complementary positions, although the priesthood were almost certainly the subordinates in any hierarchical structure.

As with reciprocally based societies, these rank-redistribution societies often continued for considerable periods at a relatively constant level of production. Many declined, even disappeared, as a consequence of environmental vicissitudes. Because of their proximity to the Malthusian limits, both needed to be constantly attempting to increase their productivity, too, but the social organization of the rank-redistribution societies ensured that they were more likely to sustain such increases than were their reciprocal predecessors. Both landlord and priesthood classes, but especially the former because of the celibacy requirement of the latter in many societies, grew through natural processes. Their power allowed them to demand more from their subjects in the form of tribute, in order to support their offspring. Success in such demands led to an increase in the size (at least in an absolute sense) of the non-productive or 'parasite' groups within the society.

To meet the demands of the expanding parasite group, subjects were required either to yield a greater proportion of their production and/or labour to the overlords or to increase their productivity. The former would have been counter-productive, in the long term, for it would almost certainly reduce the productivity of the subject class and put the whole society into a downward spiral of falling living standards. Increasing production through greater inputs also had finite limits which were easily reached, since most time would already be spent in productive activity. Only increased efficiency of productive activity would allow continued growth, of provision and thence of the parasite group. This would probably have required tools, manufactured by craftsmen and, sooner or later, raw materials not available locally. To obtain the latter, trade with other societies was necessary, so expansion of rank-redistribution societies eventually led to the establishment of trading links, between separate, probably independent societies.

Initially exchange involved bartering, with bargaining being employed to establish that, for example, x units of commodity A were equivalent in value to y of commodity B. In bartering between equals, such bargains would reflect levels of supply and demand, which undoubtedly fluctuated widely, with consequent changes of value. But if at all possible, the parasite group of one society would seek, through its military might, to dominate the others with whom it traded, so as to influence strongly, if not control, the bargaining process. Trading enterprises were often associated with military missions, therefore, leading to the establishment (usually ephemeral) of rank redistribution between as well as within societies.

Contact between societies, particularly between those separated by considerable distances and occupying different environments, usually indicated a wider basis for trade than was first appreciated. An increasing number of commodities would be either offered for exchange or demanded by the stronger in the rank-redistribution system, and more societies would be drawn into the trading orbit of each. Multilateral bargaining of commodities is unwieldy in such contexts and continued expansion of trade, and thus of benefits for at least the various parasitic groups, required the development of a common medium of exchange. The next stage in societal change, therefore, was the grafting of a system of money onto the rank-redistribution organization.

With the expansion of societies based on monetary exchange, the rank redistribution of commodities declined in importance. Instead of demanding goods and labour from all of their subjects, the landlord groups required rent for the land occupied by the producers (and, later, from those whose production was not land-based, a tax on income). The money received was used to buy the required commodities (in a buyer's market if at all possible) and to pay employees, such as the workers in the landlords' households and their military and administrative support staffs (the latter were involved in securing the collection of rents and taxes). To meet the rent demands, the subject groups had to produce and sell, to obtain money, which was also required in order to pay other dues (to the priesthood who supported the landlords, for instance) and to purchase other goods which they did not (or could not) produce for themselves. Their basic needs were for the means of reproduction and survival – food, housing, clothing and the requisite tools for maintaining and, if at all possible, expanding their productive capacity. In addition, they may have wished to buy other goods which were 'non-necessities' but which were desirable; the more that they produced and earned, the more they could hope to fulfil these desires.

A rudimentary exchange society contained three groups, therefore:

(1) a relatively small parasite group, who were the descendants of those at the peak of the rank-redistribution society which preceded it, whose power was based upon ownership of land and whose material and other requirements were met through the exercise of that power;

(2) a subject group, comprising both agriculturalists and the producers of other needed goods; and

(3) an intermediate group, employed by, or associated with, the first group – the main members were the priesthood, the military and the administrators.

The society operated with the subject group producing the requisite goods and services in order to satisfy the imperatives of their rulers and to sustain their employees.

Mercantilism

The next major stage in the evolution of modern society involved the appearance of a group whose major role was buying and selling. These merchants perceived the potential value of trade and saw it as a way of obtaining an income either without actually producing the means of subsistence and reproduction themselves, or without coming into open conflict with those who extracted those means out of an unwilling subject populace. The service offered by the merchants was that of trade articulation, thereby absolving individual sellers from the need to find buyers who wanted their produce and individual buyers from seeking out those producers with goods to sell. The bulking service provided by merchants, which involved buying up the fruits of a number of producers and transferring them in bulk to markets, was of most importance in long-distance trade, in which direct contact between producer and consumer was difficult. For local trade, in which both producer and consumer were neighbours, demands for a merchant's service were few, but as the market exchange economy expanded, and as each society specialized in that which it could produce most efficiently (given that demand existed), so the proportion of production which was locally consumed declined and long-distance trade became crucial for the maintenance of societal consumption norms.

Whereas a producer sells and then buys, thereby needing money at the second step only, a merchant buys and then sells, so requiring a sum of money before he can start; this money is generally known as capital. To begin trading as a merchant, therefore, a person must have either a store of capital sufficient to finance his first venture or access to somebody else's capital. The former situation may have been the case with some independent producers, perhaps craftsmen, who had been able to save some of their earnings, but the most likely holders of capital would have been landlords, who could also have obtained it through the rent and tax systems that they operated. Landlords, certainly, were most likely to have been the only ones able to offer loans to those wishing to establish themselves as merchants.

Whatever the source of the initial capital, it was put to use by purchasing commodities from their producers and selling them in bulk, at a greater price than was paid. The likely buyer was another merchant, probably from another society, who disposed of them to a population of consumers who lived some distance from the area of production. Out of the income from the sale, the merchant who made the initial purchases may have had a loan to repay, with interest: after this and other costs (those of transport, for example) had been met, the remainder was available for meeting his needs and desires, plus those of his dependants.

The merchant's role, therefore, was to use part of the surplus achieved in production – that money held by some (usually the landlords) which was not immediately required for the purchase of consumption goods – to generate even more; he did this by realizing the potential of trade that had not previously been articulated. Rents from land were the major source of the initial surplus, thus creating close interdependence between the landlord and merchant groups. (Such interdependence may have been accentuated by landlords who invested in mercantile ventures by their kin.) Once successful, however, mercantilism was then able to generate at least a part of its capital demands, and to invest in its own further buying for subsequent selling. Some merchant ventures provided a substantial return, much greater than was needed to satisfy even the most avaricious of consumption needs, and provided a pool of capital available to finance further mercantile ventures. This then called into being other 'merchants' whose role was to articulate the flow of capital rather than of goods: banking emerged as the occupation which collected together the surplus money, bulked it, and lent it to merchants seeking capital for trading enterprises. Not only did this increase the division of labour, it also assisted in the creation of further surpluses (for the lenders, or depositors, who were paid interest, for the bankers themselves, who charged a price for their services, and for the borrowers whose trading activities were a success) and increased the volume of available capital. (Banking of itself achieved the latter, for all moneys deposited are then credited both to the lenders and to the borrowers.)

Like reciprocal, rank-redistribution and money-exchange systems before them, mercantile societies could remain virtually stagnant, with the merchants articulating sufficient trade to satisfy the demands of their patrons. Such satisfaction was rare, however, especially as money-lending became more common and mercantile operations involved sharing the surplus (the difference between buying and selling price) among the providers of capital, the articulators of its provision and the merchant venturers themselves. In such situations merchants were obliged to offer a return on the investment made in their activities; if they did not, or if the return that they offered was less than that promised by others, they could not attract the needed capital. As long as those with capital to invest sought always to achieve the greatest return, albeit with some security as well, and thus to increase both their ability to consume and their title to further wealth and aggrandizement, then to win the investors' confidence merchants had to offer success. Concurrently, if the results of mercantile success were not recognized by conspicuous consumption, and by the desires of successful merchants to join the capitalist class, sale of the commodities involved in the trading so financed would not occur, and the merchants would not achieve the desired returns. The system was a circular one which fed off itself and demanded its own success.

Mercantile success required the merchants to buy as cheaply as possible, and to sell as expensively as possible; it also demanded that they trade in as large a volume of goods as possible. Thus, when buying from producers, the merchants aimed to manipulate prices to low levels, whereas when selling to consumers they attempted to raise prices to maximum feasible levels. This created a contradiction, however, for the producers were also consumers (though not of the goods that they produced), so that if the prices that they received were low, they could not afford to buy large quantities of other goods and thus satisfy the demands of the merchant class as a whole. A consequence of this was a great pressure on producers to increase the volume of goods offered for sale, which meant increasing their productivity, while merchants put pressure on consumers to buy more, even if this meant them borrowing money in order to afford their purchases. Both processes, investment in higher productivity and borrowing to finance purchases, involved producers raising loans which they had to repay with interest; to achieve the latter, they had to produce more (or, if they were employees rather than independent producers, to work harder).

Mercantilism, then, was a self-propelling growth system, in which the continued expansion of trade was paramount. Without such expansion, neither the merchant nor those dependent on his success (buyers, sellers and money-lenders) could maintain their economic position, let alone advance it. Many business arrangements were invented to guarantee mercantile success, for individuals or groups with common interests, rather than for the merchant class as a whole. One of the most common was the monopoly, which constrained, if not removed, competition and price-cutting. Merchants sought either or both of the sole right to buy a certain commodity in a place and the exclusive privilege of selling in a given place. Such monopolies were usually bought from the owners of the places concerned – the landlords – and were used to advance individual causes through the imposition of scarcity.

Into capitalism

Mercantilism was based on trade, on the creation of wealth by the processes of buying and selling, conducted by individuals who had no direct part to play in the productive process. In a simple market-exchange system the price paid to a producer covered the cost of his production (which may only be his labour plus certain materials) and provided his standard of living. The greater the amount of labour expended in production of the commodity, the higher the price he should get. Other contributory factors to price determination included the ratio of supply to demand. If a producer's commodity was in short supply relative to the demand for it, he could ask for, and receive, a price in excess of the cost of production, thereby providing an income which was more than sufficient to cover the cost of living for himself and for his dependants. The difference between the 'selling price' and the 'making price' (the latter is the cost of living for his family – the costs of the reproduction of labour – plus any other expenses incurred, such as rent, materials and hired labour) is the 'surplus'. In mercantilism, part of the surplus was taken by the merchants, those who articulated the trade between buyer and seller, and so ensured the selling price. In turn, part of the portion of the surplus received by the merchant went to meet his costs (including those of maintaining his dependants and of the support services which he had used in the trading venture – in transport and money-lending, for example).

In the absence of monopolies, success in the mercantile system required growth. With competition, if one producer reduced his supply to the market (via the merchants) he could not push up prices, since other producers would fill the void which he left; similarly, if he raised his prices he would not sell, as competitors would undercut him. He had to expand his production; industrial capitalism provided the means.

Industrial capitalism

Some producers in a mercantilist system owned their means of production, the sources of their livelihood: craftsmen owned the premises on which they worked, for example, and miners owned the land containing their mineral resource. Agriculturalists owned their labour power, but the land that they worked was in many societies owned by others, and their use of it depended on the producers paying certain dues to an overlord. Thus each merchant faced a mass of independent producers and sellers, and he was unable to control their production. He could only exhort them to produce more, thus yielding a greater absolute volume of surplus to him, through appeals listing what can be achieved with a greater income in terms of consumption. Industrial capitalism, however, introduced a system whereby producers could be alienated from their means of production, virtually required to produce as much as possible, and dictated to with regard to the proportion of the selling price that would be returned to them (although the latter clearly had to cover the costs of the reproduction of labour, if productivity was not to fall).

Under this new system, money was not invested in the fruits of production, as with the purchases made by merchants, but in the process of production itself; capital was invested in labour, not in goods. Those with capital, such as landlords and successful merchants, or those with access to it, usually from the same economic groups, could gain control of the means of production – the raw materials. Landlords, in particular, could deny access to land, and thus to sources of food and other raw materials, among their tenants. Lacking any ability to produce the necessary wherewithal for the reproduction of their labour power, the alienated workers could then be required by those with capital to work on the available materials, thereby earning the necessary income for the purchase of subsistence. They had to sell their labour. They could bargain with those seeking to buy it, but the latter were much stronger. Without work, the alienated labourers could not survive; without workers, the capitalists could not produce goods for sale. In the short term, however, whereas the latter's capital resources would support them in a period of no income, the former had no such reserves on which to call.

Under industrial capitalism, therefore, an increasing proportion of capital was employed to put people to work, rather than to purchase the products of labour. Employing labour was not novel, of course, for many, if not most, craftsmen employed assistants (who eventually succeeded them); the difference was that under industrial capitalism the scale of employment expanded and the employer no longer worked at the same tasks as, and alongside, his employees. With the new system, production was governed by the desires and whims of capitalists who invested in putting people to work and taking the surplus, the difference between the 'making price' and the 'selling price', as the reward for their investment and initiative. Manipulation of this surplus, relative to the prices of commodities and the costs of reproduction of labour, allowed the capitalists to control their employees and to demand increased effort from them. In reciprocal societies the social structure was egalitarian, because of an equal contribution to production by all and an absence of any economic division of labour. The progress through rank redistribution and mercantile capitalism to industrial capitalism increased the level of inequality within society, with the result in the last of these systems being that the mass of alienated labour was made subservient to the demands of the minority, the owners of capital.

Small-scale industrialism, with most producers as employees (in agriculture as in other economic sectors), could stagnate, with no increases in production, a stable population, and the labour force producing just sufficient to satisfy capitalists' demands. But such stagnation was rare. Growth was and is the driving ethos of industrial capitalism, not least because of the usual pressing demands of increasing populations, which stimulate both consumption and wage-requirements, and the desires of capitalists for greater profits.

Some growth in the surplus achieved by the capitalists can be obtained by ensuring that the labour force either works harder or accepts lower wages, thus increasing the gap between the 'making price' and the 'selling price'. As described earlier, however, both solutions have finite limits in the short term, and their use is likely to prove counter-productive. The solution is to increase labour productivity by reducing the 'making price' without lowering the wages of the workers; they must make more with the same amount of effort. (The exercise of monopoly, by which the 'selling price' is manipulated, is also a feasible strategy.) To feed the desires of capitalists for good returns from their investments in labour, and to allow them to attract investments and loans from others, industrial capitalism was organized and reorganized to improve levels of productivity. Initially, much early industrial capitalism involved the industrialists providing materials, and perhaps simple tools, to their employees, who worked on them at their homes. The improvement of productivity required that their output per unit of time be increased. To achieve this it was necessary to provide better tools – machines – and to ensure greater oversight of their work: together, these two solutions led to the introduction and development of the factory system, by which employees were brought together to work in a single building under the control of the capitalist (or his administrators), using machinery which he had installed. But although industrial capitalism is inextricably bound up with the factory system, which is its most obvious outward manifestation, its distinguishing feature has been the alienation of labour power which enforces the working for wages.

As with other forms of capitalism, industrial capitalism has had an inherent drive for growth. More must be produced, and then sold, to ensure the needed returns on investment. Again, this leads to the contradictory situation whereby forcing down wages so as to increase returns on investment (profits), while at the same time attempting to maintain if not increase prices, produces a reduction in consumption and thus capitalist failure. Wages must rise, to keep pace with prices, in the system as a whole, and this could only be ensured by growth – more production, more productivity, more consumers, more people. The resulting greater profits lead to a greater pool of investment capital, seeking the best returns; to attract this investment, the productivity potential of workers is increased by improving the tools with which they work. Better and more complex machinery means more production, more and bigger factories, and more goods seeking purchasers. Growth demands more growth.

The industrial capitalist system comprises many parts, individual firms competing with each other to make and to sell more, to achieve greater profit levels, to provide better returns on investment, to attract more investment, to make and to sell more . . . Some of these firms have failed as a consequence of the success of others. Indeed the success of the system as a whole depends on there being failures as well as successes, and on the re-channelling of investment towards the latter. Out of this re-channelling has emerged the present stage – late capitalism.

Late capitalism

Industrial capitalism is characterized by a large number of firms in each sector of the economy (not only in manufacturing but also in agriculture and in the service sector); all compete for sales and for investment; none has a monopoly. Some of the firms are more successful than others, which provides them with a base for expanding both production and productivity, but success in these requires an expanding market. This may be provided by either or both of population growth and growth in real incomes (generated by in creased productivity in most sectors) in the traditional market, but new consumers may have to be sought. The latter can be achieved via improvements in transport technology (as detailed later in this book), giving access to formerly unserved populations. Eventually, however, the potential market for the product is saturated, and the only way in which a firm can expand its sales is by capturing the market (or at least part of it) served by its competitors. The more successful, and therefore more profitable, firm is enabled to do this because its profits can be used (along with relatively easily attracted other investment) either to purchase a competitor who, because of poor performance, lacks capital, or to subsidize price-cutting competition which forces weaker firms to yield their market share, by either selling or collapsing.

Within any sector of an economy, therefore, inability to expand the market will lead to a process of concentration, whereby competitors are successively excluded as the weaker firms disappear. Eventually it will be dominated by only a few large firms, and perhaps even by a single, monopolistic producer. In the former case, the few remaining firms may continue to compete and to seek to destroy their rivals. The perils of potential failure are great, however, and firms are quite likely to settle for a relatively stable division of the market between themselves, perhaps even colluding over pricing and other practices.

Once either a monopolistic or an oligopolistic situation has been established, the successful firms will face difficulties. If the market is fixed, then the only potential for increased profits lies in the manipulation of the difference between the 'making price' and the 'selling price'. A major problem with increasing productivity, however, concerns the disposal of the extra output. Investment will not be attracted, although the firms may require it in order to replace obsolete plant and to retain their competitive positions *vis-à-vis* others in the sector. More importantly, there will be no apparent outlet for the accumulating profit income. Market saturation under either oligopoly or monopoly creates a crisis for investment.

The likely solution to this crisis is for capital to flow into those sectors of the economy where the market is still expanding and potential returns are relatively high. Thus a successful large firm in a saturated sector will seek to invest its profits in, if not take over, firms in other sectors. These may be firms in sectors associated with its own – suppliers of component parts, for example, or customers for its products. But if such investment opportunities do not exist, then the firms with capital to invest will seek to place it wherever potential profits are attractive, irrespective of whether the invested-in sector has any functional relationship to the sector from whence the capital comes. Thus concentration within particular sectors is followed by centralization, by firms extending their control over several sectors, so that whole sections of the economy come under single ownership.

As more and more sectors become saturated, and are unable to expand their markets, so the problems of late capitalism expand. Investment can raise productivity, but because of the lack of demand it is difficult to sell all of the output, and the consequence is unemployment for those workers replaced by machines. Investment capital finds it difficult to locate attractive destinations, and many of the large firms are close to failure; a potential consequence of the latter is large-scale unemployment and the failure to produce needed goods, because of a lack of market growth, whereas as a result of the former capital seeks investment in non-productive sectors (such as antiques and works of art whose value, it is assumed, will remain constant, if not increase). Profits from the latter are usually short-lived, if the non-productive investments prove to give little or no return (which is probable as increasing volumes of capital flow towards them), and firms continue to face the spectre of failure.

No society has reached the ultimate extension of this process of centralization, which would be a single firm, in part because some sectors are not attractive for investment even when there are few alternatives; production in them is best continued by small firms. Nor is concentration so extensive that most sectors in each society are controlled by a monopoly. But there are many firms whose investments are spread across a large number of sectors and which are experiencing problems of maintaining profit levels, of continuing production and of expanding markets. Many are faced with the need to produce less than their capacity would allow and find that they do not need their entire labour force. Lay-offs lead to reduced consumption, however, and initiate a downward spiral of declining demand.

To counter these difficulties in late capitalist societies, the state has been built up as a support mechanism. This involves it in a variety of roles. To stimulate production, for example, it may create demand by 'manufacturing money' in the form of a national debt which is owed to itself (if outside loans are not available), and this money can be used to purchase goods and maintain industrial profitability. Alternatively, it may use its income, gained from all residents and not just the capitalists, to subsidize capitalist ventures, by paying producers the difference between the price that the market will bear for their products and the necessary price which would give them their desired rate of return on investment; this will keep people in work and ensure the circulation of money and its expenditure on goods and services. Further, the state may take over an unsuccessful firm (nationalize it) and keep it in production despite its lack of profitability, for the same ends; people are kept in work by paying their own wages. (Because money circulates, and one person's wages create the demand which generates wages for others, each employee of a nationalized industry pays only a portion of his or her own wages.) Finally, it may create markets for goods which otherwise would not be bought. These may be in other societies, through the granting of foreign aid, money allocated to other societies with the caveat that it must be spent either in the donor country or on its products. Or they may be for goods not purchased by individuals, of which the major examples are armaments. These may never be used, except in training, but if all countries are doing the same, and each is convinced that peace and the 'balance of power' can only be achieved by a continual 'improvement' in weapons, then there is an infinitely extendable market for such goods (unless concentration and centralization lead to the amalgamation of countries).

Such state activity has to be paid for within the capitalist mode. In part the costs can be met by the profits of the successful sectors of the economy (presuming that there are some) which have buoyant markets. In part they can be met by creating money, although this is inflationary and, in the end, self-defeating. And in part it can be paid for by all, since a portion of the 'making price' (the wages received by labour) is necessary for meeting tax and related demands; the state must ensure that money is used wisely, so that its expenditure generates income through the multiplier processes outlined in later chapters, and the portion which must be taken in taxes is kept low. Thus the role of the state includes absorbing part of the cost of making profits by private capital, thus reducing the 'making price' relative to the 'selling price'. But as the profit-making function of capitalist enterprise is founded on growth, so the state is called upon to absorb an increased proportion of the 'making price'. Its expenditure must increase more rapidly than it can increase its revenue, because it must allow capitalist profits. Thus late capitalism is also characterized by an increasing fiscal crisis for the state.

Finally in this section, two other features of late capitalism remain to be stressed. The first concerns the occupational structure of the workforce, which becomes increasingly dominated by non-productive activities (what are widely known as the tertiary – notably retailing and other commercial activities – and quaternary – such as research and development – sectors). This domination is the current culmination of a sequence which began with the movement of some individuals away from food production in the rank-redistribution societies, as agricultural productivity increased and a few could be released from providing their own means of subsistence. Indeed, the whole of the developmental sequence outlined here depends on ever-increasing productivity in agriculture (and the other sectors providing raw materials from environmental sources). Initially, this resulted in a growth in the number of people involved in trade and related activities and then, with the onset of industrial capitalism, a rapid increase in the proportion of the workforce engaged in manufacturing. As industrial capitalism became more complex, with the need for finance and trade on a large scale, the manufacturing sector began to decline relatively and the tertiary and quaternary sectors came to dominate the occupational structure. This was occasioned by the great increases in productivity within manufacturing, which released large numbers from the need to be involved in production; the majority of workers in late capitalist economies are supported by a portion of the difference between the 'making price' and the 'selling price', for they are in no way involved directly in the manufacturing processes.

The second feature concerns the cyclical nature of the fiscal crises outlined above. In any one industry, once its market is saturated, growth is impossible and investment capital flows elsewhere. As a consequence, the industry, as it declines in profitability, becomes increasingly backward – its machinery becomes obsolete and its technology outmoded. In a highly developed industrial system, decline in one industry is associated with decline in many others, because of the interdependence which is a major characteristic of this stage of industrialization. Most industries produce not for the final consumer but for other industries, providing the machinery and other technology used in the vast myriad of manufacturing processes and creating the components which are fabricated elsewhere into final products. Thus a decline in one will have almost immediate effects on those it is linked with: if shoe factories are in decline, those which make their machinery are likely to decline as well, along with the suppliers of leather, rubber and other raw materials used. As a consequence, in late capitalism much more than in industrial capitalism, a crisis of profitability is likely to affect the whole economy and not just particular sectors of it. Eventually, according to some economic theorists, the crisis will pass, as production falls so far that profit expectations are reduced and, with increasingly unsatisfied markets, the possibilities of growth re-emerge. A slump should be followed by a boom, as indeed was the case under industrial capitalism, sometimes in certain sectors only and sometimes in whole economies. But in late capitalism, because of concentration and centralization, the potential of a slump is catastrophic, so that the state has had to accept the role of stabilizing the economy and ensuring that such downturns, consequent upon reduced profitability and attractiveness to investors, do not occur. The state is the regulator of the late capitalist economy, protecting the large firms from the vicissitudes of trade and profitability cycles, and thus ensuring the jobs and livelihoods of the mass of the workforce, including those in the tertiary and quaternary sectors, who depend on the viability of the late capitalist monopolies and oligopolies.

Classes and the capitalist systems

The egalitarian, reciprocity-based societies described earlier were largely communistic in their operation. All property, with the exception of personal effects and, perhaps, dwellings, was communally owned; land was communally worked; and hunting/gathering expeditions were communally organized, as were conflicts with neighbouring societies (which usually occurred when one or the other was weak and suffering from environmental stress). There were no differences of rank, status or wealth, and no positions of power and influence to be protected. This has not been the case with any of the later systems, however, in all of which an economic division of labour has been accompanied by a social division of society.

At their initial stages of development, rank-redistribution societies contained two major groups – the elite, which was small in number, and the producers, who provided tribute and the means of subsistence for the élite. The former occupied a privileged position, one which they sought to protect and to ensure for their descendants. To achieve this protection, barriers were erected which prevented producers entering, or even aspiring to, membership of the elite. To reinforce these barriers other groups – military, administrative and priestly – were employed by the elite to ensure their own separateness. These new groups occupied intermediate positions and status. They were neither producers nor élite; they owed their positions, and a way of life usually slightly superior to that of the producers (in order to buy their loyalty), to the élite, and if the latter lost power they would lose status and their comfortable existence with them.

As money-exchange societies developed, so the division of labour proceeded, to be followed by a more complex social division. The élite maintained their status by keeping for themselves ownership of the major means of production, land; all others were allowed at most only the use of sufficient land to provide for their basic needs, and perhaps a small surplus for trading. Ownership of land conferred status, and the majority of the land was retained by a small minority. The new intermediate groups, merchants and craftsmen, were also as rigidly defined as those of the elite, and entry was restricted to those from certain social origins only, usually the kin of those already in the group, so that occupational status was inherited. Thus such societies comprised a set of discrete components based on occupation, power and status (the last two were consequences of the first); for most members of each society, the group into which they were born was that in which they lived and died, and movement from one class to another was rare, if not proscribed.

Extension of the money-exchange system into mercantilism, and thence industrial and late capitalism, required some reduction in intergroup rigidity; social boundaries had to become somewhat porous to facilitate the economic processes. Mercantilism and capitalism have been built on growth, which requires the creation of new occupations, plus the growth of many more and the decline of a few others. Occupational mobility, both intra- and inter-generational, was necessary to fuel these changes. And yet, at the same time, those occupying the more privileged and better-rewarded positions of power and status did not wish their own niches invaded and their privileges placed at risk. Those at the apex of the system, the landowners and the capitalists, were relatively secure, for their monopoly over land, trade and, later, manufacturing ensured their control over the means of production and prevented others from amassing large volumes of capital; as trade and then industry expanded, the balance within the élite shifted, with the landlords suffering some decline in status because their controllable assets were fixed.

Below the élite came those who, in the capitalist system, were required to sell their labour power for a wage. This working class has always been far from unitary, however. Indeed, a new group, often termed the middle class, developed within it as the skilled servants of the capitalists, those whose expertise was necessary to the profit-making demands of investment. For their services and loyalty to the capitalist cause they have been relatively well rewarded, compared to those involved in production (the 'real' working class). Next in line for status and rewards came the skilled craftsmen, and below them were the performers of the many relatively unskilled tasks necessary for successful production.

The growth of mercantilist and capitalist systems offered many opportunities for economic and social advancement, especially for those with inherited or learned skills. Incorporation of new territories into a trading system, for example, allowed others to join the landowning class, but a much more important stimulus to economic and social mobility was the rapid growth of the industrial capitalist system. This made great demands for skilled workers, to invent and build the machinery necessary for increased productivity and profits, to invent the new products which would be demanded when the market for others became saturated, to organize the increasingly large volumes of goods traded and of capital invested, and to train those who would undertake all these tasks in future generations. The potential for movement across social barriers was expanded manyfold. Members of each group were encouraged to aspire, for themselves, but even more so for their children, to positions in the more prestigious 'higher' economic and social strata, which offered greater monetary and psychological rewards. They might even, especially in very rapidly expanding societies experiencing acute shortages of enterprise and labour, be encouraged to aspire to élite status, to join the capitalist group, although these barriers were only substantially lowered when the potential for profit-making far exceeded the capability of the existing capitalists to realize it.

Industrial capitalism comprises two main classes: the owners of capital, who invest in labour in order to produce a surplus; and the owners of labour, who sell their skills to the capitalists in order to earn wages which are used to buy the means of subsistence and reproduction. The latter is by far the larger, and, as already indicated, is far from homogeneous. Its major division is between producers and non-producers, those involved in the manufacture of goods for sale and those involved in the facilitation of both manufacture and sale; the latter are generally known as white-collar workers, and the former as their blue-collar counterparts. With the movement into late capitalism, and the saturation of many markets for the production and sale of 'necessities', a further group has emerged, of non-producers who provide services (in leisure and entertainment, for example) to those with income remaining after they have ensured their subsistence. Each of these major divisions comprises many minor subdivisions, based on such criteria as type of work done, the amount of skill that it requires, and even the employer. The capitalist class is more homogeneous, but different groups within it compete for control over the means of production and the ingredients for financial success. (A major contest in most societies has been that between agriculture and manufacturing.)

Capitalist societies are characterized by conflict between these two classes, between the capitalists who wish to appropriate as large a surplus to themselves as possible, and all others who are seeking as large a portion of the 'selling price' as possible. Such conflict occasionally erupts into confrontation, notably during periods of capitalist crisis when the absolute size of the surplus is small. Usually the contests are unequal, however, for control of the means of production by the capitalist class allows it to deny work to the others, thereby immediately affecting their ability to ensure their own subsistence. Further, internecine conflict within the working class is often encouraged to divert attention from the major issue. Various occupational and industrial groups compete one with another for rewards and status, protecting differentials and contesting anomalies. All such contests must be settled eventually by compromise, in which one group is the loser. In many societies the importation of new groups from other societies to provide needed labour during a period of growth has fuelled latent xenophobia and has provided a 'scapegoat' group who suffer first and most in periods of capitalist crisis.

Concentration and centralization in the capitalist world are paralleled by similar trends in the working and middle classes, as a consequence of the small number of firms remaining and the large size of their workforces. This somewhat reduces the intra-class conflict and allows workers to develop a more unified and stronger bargaining position relative to that of their employers, through combination in trade unions. Such large unions often themselves become quasi-capitalist in their organization, however, amassing funds based on small levies on members in order to finance their activities. They then must invest these funds, to ensure their continued viability and growth (especially in periods of inflation); pension funds, in particular, are invested in capitalist success (i.e., achieving satisfactory profit-levels). By making such investments the workers help to legitimize the system with which they are in conflict: their vested interest in its continued success encourages them not to take action which would threaten its stability. Similarly, individuals are encouraged to invest their savings, however small, in capitalist enterprises and in capitalist activities such as home ownership, thereby encouraging their commitment to the system's success.

Industrial and late capitalism are characterized by an economic and social division of the population into classes, therefore, distinguished by their position relative to the means of production and, as a consequence, their power to control societal events. The larger class, in particular, is divided into a number of competing groups, each both in conflict with the capitalist class over the relative divisions of the fruits of capitalist production and competing with other groups for relative status, power and rewards. Encouragement of the latter contests by the capitalist class, coupled with minor concessions to the working class when necessary, or to portions of it, ensures maintenance of the basic set of economic and social relationships.

The state and capitalism

The élite in rank-redistribution and money-exchange societies – basically the landowning class – require an apparatus to protect their position. In small, locally circumscribed societies this apparatus may be slight, but any expansion of a society will require its growth too. Much of this expansion has involved either the incorporation of new territory, formerly occupied at lower levels of economic development, and establishment there of a subsidiary landed class owing loyalty to the established élite and paying them dues, or conquest of another society, making its landed interests subservient and liable to dues. In either case, military and administrative cadres are needed, employees of the élite, landed class who will protect its dominant status and collect the tribute that it demands.

This administrative and military apparatus is the nascent state, established by the dominant class to protect its interests. As the economic system progresses through the various stages outlined previously in this chapter, so the demands on this apparatus grow, to protect trade, to guarantee monopolies, to govern exchange mechanisms, and so on. For long periods acceptance of the legitimacy of the economic organization will leave the military in a minor role, with coercion of the working class being substituted by their incorporation into the monetary system and then onto the fringes of the capitalist class itself. The only role for the military, except a psychological one created by their very presence, will then be to protect the society from other societies, all of which may be seeking to expand the sphere of influence of their capitalist interests.

Although necessary to mercantilism, the state played a minor role in that economic system, and was not divorced from the established élite. Its major functions were defence of the society's interests against outsiders (a military role), maintenance of law and order and the protection of private property (a police activity), and provision of the means for the proper conduct of business (a civil code and a legal mechanism). These remained with the growth of industrial capitalism, as, for example, each society's manufacturers demanded protection against their competitors in certain markets and encouragement in others. Increasingly, the state took on more civil functions involving the protection of capitalist interests; it eventually evolved as a separate sub-class itself, appointed (or elected) by the capitalists (property-owners) and employing members of the middle class, already closely allied to capitalism and its aims.

With the progress of industrial capitalism, so the state has become increasingly involved in mediating the conflicts between the capitalist and working classes. Such mediation often favours the former, not only because the state very largely represents it but also because of the general ethos that what is good for capitalism (growth and profits) is best for all members of society – without capitalist expansion, the ethos runs, wages cannot be paid. Some concessions are made to the working class to ensure their legitimation of the system, such as extension of the franchise. This could allow them to dominate the state apparatus, but such a trend is countered by the incorporation of the working class into petty capitalism and by ensuring that the state mechanism remains largely in the hands of the educated classes, the allies of capitalism. In this way, even an elected government from the working class, which will be supported by a bureaucracy from the middle class, is likely to accept the need for a healthy capitalist economy.

During late capitalism governments have introduced concessions to the working class in order to secure their support for the economic and social system. Many of these comprise what is widely known as the welfare state, and include expenditure on health care, at least for the young and the old, on education, on housing, on unemployment benefits and on various types of pension arrangement. Although these are paid for out of the surplus, at least in part, because taxation affects all classes, they are not widely perceived as deleterious by the capitalist class. Education and health services are necessary to ensure the reproduction of a labour force which can perform the tasks required of it, and their provision by the state subsidizes capitalism to the extent that their total cost is not met out of profits. Similarly, the so-called Keynesian policies of economic management, aimed at smoothing out the booms and slumps of capitalist business cycles, are also in the interests of capitalists, while they are often introduced to protect the working class from the alternative perils of unemployment and inflation. Thus the liberal-democratic welfare state serves capitalist interests while at the same time mollifying the working class.

Under late capitalism the state becomes a major protector of the often ailing capitalist body. It intervenes by attempting to regulate the economy so that profitability is ensured, it protects monopolies and oligopolies from outside competition, it subsidizes ailing firms, it provides the infrastructure for capitalist operation, it referees conflicts with the working class and enacts many other policies aimed at encouraging investment and growth, while at the same time protecting employment. In doing all of these things, it commonly faces contradictory pressures, from different groups within the capitalist class, for example, as well as within the working class, and it must decide between the conflicting demands made of it. Increasingly, too, it faces the problem that the largest firms within the capitalist class are now organized multinationally, putting it in conflict with other states in terms of competition for the benefits of growth in those firms; such international conflict may force it to adopt policies contrary to those which it would promote if all firms were its own nationals. Finally, the size of the state apparatus, in particular its large bureaucracy, is itself the creator of new pressures and conflicts, for the bureaucrats act to protect their own positions and to ensure policies that will promote their own economic and social aggrandizement.

The state is a critical instrument in all types of economic system after the reciprocal, therefore. In the early stages its role is a simple one, to protect the interests of the élite, but under capitalism its position within the society's operation becomes both more central and more crucial. It occupies a position between the two main conflicting classes and must act to ensure that a major crisis is never precipitated that might destroy everybody's interest in the *status quo*. Finally, in the late capitalism stage it is increasingly a major customer, or stimulator of custom, whose purchases are necessary for capitalist survival. It is the recipient of many competing claims, and different government personnel may favour different sections of capital and of the working class, but the end remains the same: to ensure the continued healthy operation of capitalism.

Alternatives to industrial and late capitalism

Capitalist economic systems are based on inequalities, both economic and social, since one relatively small class has a virtual monopoly over crucial economic decision-making and so can manipulate the political and social structures to its own ends. Liberal democracies have in many cases produced governments whose aims have been to redistribute economic power and so create a more egalitarian distribution of income and wealth. In most cases they have succeeded in improving the absolute conditions of the working class (as regards housing, schooling and health, for example), but they have had little impact on their position relative to that of the capitalist class, since every step forward for them is matched by a further step taken by the capitalist and 'middle' classes, involving the erection of new barriers to entry into high-status and reward positions. (Although such absolute improvements can be credited to liberal-democratic governments, it is quite likely that they would have been produced by the capitalist class in any case, in order to increase levels of consumption, to avoid overproduction and to ensure reproduction of the needed labour force.)

Mild liberal-democratic changes of this type characterize much of the industrial capitalist and late capitalist world. Elsewhere, attempts have been made to replace capitalism, with its ethos of private property and control by a few, with a system more akin to that of the early egalitarian, reciprocity-based societies, which would nevertheless ensure levels of living comparable to those achieved under late capitalism but with different consumer orientations.

The basis for these alternative societies, most of which have been created out of either mercantile or industrial capitalist origins, is the socialization of the means of production, distribution and exchange. Private property, other than personal effects, is not allowed, and all production and trade is organized by the state, so that decisions are made for the general good, not that of a particular individual, firm or interest group. The society is organized hierarchically, with local decisions made locally, those involving regional interests being made regionally, and so on.

In essence such a decision-making structure is quasi-anarchic, since it allows considerable autonomy to each local cell. In effect this usually breaks down, for two reasons. The first is that, since industrial growth is dependent on large-scale production and organization, the local scale is not apt for societies which aspire to material standards well above those of minimum subsistence. A large, centralized bureaucracy is needed to organize and operate a complex industrial economy which lacks individual, independent decision-making units (firms), and so the system must take on a structure analogous to that of late capitalism, but without the profit motive and separation of state and industry. Secondly, political power in such a system can be as great as that enjoyed under capitalism, if not greater, since there are no groups of similar size competing with the state. Such power can be craved and schemed for by individuals. The result is a permanent bureaucratic class, headed by a trained administrative cadre; operation of a complex economy requires more than popularly elected local officials. These administrators compete for status and power within the bureaucracy, just like the state bureaucrats in liberal-democratic societies, and seek to control the whole, or large parts, of the economic machine. Once in positions of power, they can then allocate privileges and rewards to themselves and their dependants, and structure the social system in such a way that those privileges and rewards, and access to positions of influence and power, are handed on to their descendants. The aims of the bureaucrats are very similar to those of capitalists: preservation of the species. The result is a class system.

Egalitarian, socialist societies are based on communal ownership and decision-making, but their communistic form conflicts with their goal of material advancement. The result is a command society, in which most decisions are made by a highly centralized, administrative élite. In this structure labour is usually as alienated as it is under capitalism, especially where, as seems common, the élite acts dictatorially, if not tyrannically, against any opposition. The degree of centralization reflects the level of capitalist-like industrialization required; where this is considerable, the result is state capitalism.

Summary

As pointed out before, the purpose of this chapter is to provide a context for the contents of the remainder of the book. This has meant the presentation of a framework only, with a complete absence of detail, so that the outline provided both lacks specifics and is little more than a cartoon of what is, in effect, the entire span of human economic and social history. The remainder of the book takes its main themes, of transition through mercantilism to capitalism, of concentration and centralization, of class conflict and of state intervention, and places the urban phenomenon within them, thereby illustrating how urban form, at various scales, reflects the operation of the processes outlined here, and how society and urban form interact in the continuing processes of economic and social change.

3. The Evolution of Urban Systems: I Pre-Industrial

A notable feature of most settlement patterns is their domination by urban places. Some of the towns may be relatively new foundations, but most are the contemporary occupants of sites which have supported urban life for a long period. To understand this continuity of settlement, and its contribution to the present scene, a historical explanation is required. The current chapter offers the first part of such an explanation, focusing on the evolution of urban places under pre-industrial conditions, which were removed from some parts of the world more than a century ago but still exist in several others. Most of the essay presents a scenario, a general discussion of the processes and patterns occurring in pre-industrial conditions, rather than a detailed description of particular places at certain times.

Urbanization under rank redistribution

Assume an area of settled agriculturalists, in which each household has its own land from which it obtains the raw materials necessary to provide food, shelter and clothing. There is some exchange of commodities with neighbours, but no trade with other settlements. Self-sufficiency and subsistence are the characteristic features of society and economy; they demand hard work and produce a low standard of living. There is, however, the potential for at least one household to develop the sort of parasitic status outlined in the previous chapter, by one or more members establishing themselves, usually through some form of religious persuasion, as deserving of tribute which will release them from the necessity to work and provide their own means of subsistence.

In such a permanently settled society the parasite household would establish itself at a central site, to which the needed tribute would be brought. This could in no sense be defined as an urban place, using the term as it is generally employed today, but it would provide the foundation for urban growth. (Such a central site could also be established in a territory occupied by nomadic or semi-nomadic peoples, but this is much less likely.) If the demands of the parasite household grew, and its entourage expanded, so the progress towards urbanization would commence.

Urban origins

Assume now that the parasite household not only wishes to increase its consumption, but also wishes to display its status and to provide a visual statement of its power over the food-producers. An ostentatious building – a temple – would clearly identify its separate and superior position within society; its construction may require the services of other non-food-producers, as might its later upkeep. Within it, the display of power and status may demand furnishings and rituals which need craftsmen to produce the former and acolytes to perform the latter, and those exercising the power may wish a more opulent lifestyle for themselves. All of these tasks will involve the growth of a non-food-producing class of workers, dependent on the parasitic élite for their livelihood, but in turn being reliant, along with the élite, on continued tribute from the food-producers. And, of course, this new class will need homes too, which must be constructed by labourers who cannot be involved in the food and raw material production.

Expansion of the society requires the development of new occupations, therefore, which means a greater volume of food and other materials (extra to their own consumption) being obtained from the producers. This additional production may be achieved through insistence on greater productivity, perhaps the consequence of harder work; productivity increases may be obtained through improvement in the tools used, however, or in the 'raw materials' of agriculture (soil, plants and animals). All three of these solutions to the problem of increasing productivity will have required more non-food-producers: administrators and, perhaps, an army to oversee the harder work (their actions may have been accompanied by the élite taking to itself the ownership of land) in the first; craftsmen to create the tools in the second, and also, probably, miners and others to provide the raw materials; and 'researchers' to develop the new strains and the new technology (notably irrigation) in the third. Thus the demands for more production are reflected in the urban node as well as in the countryside, and continued growth of the society, to meet the never-satisfied demands of an expanding élite and its associates, leads to self-propelling urban growth.

Once operational, therefore, a rank-redistribution society will be centred on a growing urban node. Societal failure, probably through an inability to obtain the needed food and other materials, will lead to urban stagnation, and possibly first decay and then disappearance. Success, however, will feed on itself, with the élite demanding more ostentatious displays of its power and status through new building programmes and creating the bases for both increases in productivity by research and development and the initiation of long-distance exchange of rare commodities between societies. Such success requires favourable environmental conditions, where the potential production of the soil is great, especially with technological improvements; this potential must also be constant, and the soil must give a good yield each season, for the ability to store will be limited. Beyond a very primitive stage, therefore, expansion of rank-redistribution societies and their towns could only proceed in the environments most conducive to increased demands for production and where improvement, by irrigation and other methods, was easy: deltaic areas in temperate and sub-tropical zones, along with fertile alluvial valleys, provided the most promising sites.

Territorial expansion and urban hierarchies

As pointed out in the previous chapter, there is a law of diminishing returns to the productivity of both labour and land, which can only be surmounted by technological improvements. Relatively primitive rank-redistribution societies would be unable to free themselves from the operation of this law, and so continued growth would depend on the ability to identify and then to capitalize on other resources of land and labour. This would require extension of the territory controlled by the rulers of the society, through policies of colonial expansion. The populations of adjacent societies might be persuaded to become part of the rank-redistribution economy; more likely, they would have to be coerced into it, by military conquest and then oversight. In some cases the adjacent lands might be empty of agriculturalists, in which case the élite would have to encourage their settlement.

To enable successful colonial activity, the structure of the society would need to be reorganized and several new functions created. Among the latter, the most important in the new areas would be administration – both civil and military – probably accompanied by religious colonization of the new subject population; to ensure contact between the colonies and the 'heartland', and the movement of the surplus production back to the élite centre, a transport infrastructure would have to be established and maintained, and the ability to move substantial quantities of goods created. All these would involve jobs for non-producers, who would need to be supported out of the production of the colony. In turn, the occupants of these new jobs would make demands on other non-agriculturalists: the military, for example, would need weapons and the transport workers vehicles. Only part of the additional surplus production obtained through colonial aggrandizement would be available to the original élite in the urban centre of the growing 'empire', therefore; the remainder would be needed to support those who established and maintained their power over the new areas, plus those who provided the means of exercising that power.

Small-scale colonial expansion, involving a few square kilometres only, can be organized from the one centre, and controlled by the single élite body. But expansion beyond easy reach of the main urban centre (perhaps a day's journey would mark the maximum extent) would demand the establishment of secondary settlements, to act as the nodes for parts of the controlled territory, as intermediate centres in the flow of demands from élite to producers and of goods in return. Each of these subsidiary settlements would have its own administrative, military and religious functionaries, with one of them of higher status; all would be subservient to the élite in the centre, but would have some independence of action in order to ensure that the latter's demands are met. The loyalty of those individuals, particularly those with the superior power in each place, must be held by the central élite, and probably would have to be 'bought' by the granting of various powers – control of the land resource would be the most likely price. As long as the 'local élite' ensured that the requirements of the central élite were met, they might be free to organize and rule their territory as they pleased.

Each of these subsidiary settlements would develop as a microcosm of the central urban node, therefore, with its assemblage of buildings and functions reflecting the power, status and roles of its inhabitants. Because some of the surplus from its hinterland was transferred to the main centre, however, this new settlement would not come to challenge the latter in terms of size and importance; part of the potential base for its growth was being channelled elsewhere. At the same time the establishment and growth of these colonial offshoots would be reflected in the central node. Control of the empire would generate new functions – in the organization of trade, for example – and perhaps lead to a division of labour within the élite group.

If the growth impetus was maintained, the empire would have to be continually enlarged, with the conquest of territory and populations ever more distant from the central node. The greater the distance between centre and colony, the greater the independence that would have to be given to the representatives of the élite in the latter, because of the difficulties of maintaining close oversight of their activities. Thus the local élite, either one established there by the élite in the centre (perhaps their kin) or one already in existence and whose loyalty has been transferred (forcibly in most cases), would have to be given considerable independence and the rights to a greater portion of the local surplus than were allotted to the sub-élite in adjacent colonies. This would allow the colonial rulers to build up sizeable urban settlements of their own, although because some tribute was being sent to the centre these would not be as large as the main node, which was drawing on a much larger area.

These processes create a hierarchy of settlements, an idealized representation of which is provided in Figure 3.1. At the apex is the town housing the élite group. The area which they control directly is bounded by the solid line. Within it is a series of six small settlements, which act

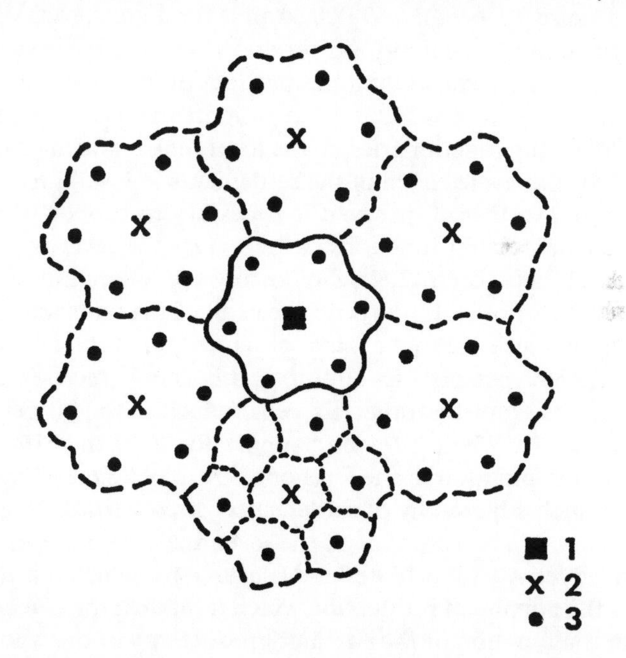

Figure 3.1 An idealized system of settlements organized into a hierarchy of three levels. The cell bounded by the solid line is governed from the level one centre, which has the largest population. Each of the six colonial cells, bounded by dashed lines, is governed from a level two centre, and within each cell local centres (level three) are used to control sub-cells, as indicated in one of the colonies.

as the foci for segments of the hinterland and which house the functionaries who carry out the élite's demands. Beyond that hinterland is a series of six colonies. Each has its own major town, at level two in the hierarchy, plus a series of smaller settlements which act as the organizing nodes for their local portions of the colonized area: the former house all the major colonial functions, including the local élite, who are subjects of the main élite group in the level one centre.

Each of these settlements shown in Figure 3.1 is the central place for a small area, and there are seven such areas within each cell. The seven largest centres (the six in level two plus the one in level one) are also the control centres for a larger region, comprising the hinterlands of seven smaller places, one of which is their own. Finally, the level one centre stands apart as the control centre for the whole system.

It is possible to estimate the relative populations of the centres in the three levels of the hierarchy. Assume that each of the seven cells of the system in Figure 3.1 produces 7,000 surplus food units, each of which is sufficient to support one non-food-producer; each of the seven towns in the cell is the centre controlling the production of 1,000 units. Of this production, 20 per cent is required to support the local élite, plus their various supporting functionaries, in the local centre, which gives a population of 200 persons for each of the settlements in level three. In the six colonial cells a further 40 per cent is necessary to support the colonial élite and its supporting functions, which gives the level two centres a population of 3,000 each (2,800 supporting the colonial functions and 200 the local functions). In the central cell all of the units not retained in the level three centres go to the main settlement, providing the means for supporting 5,800 persons. In addition, this centre receives all of the remaining units from the other six cells, which is the 40 per cent not consumed locally. These provide support for a further 16,800 units, giving a total population in the level one centre of 22,600. Thus the size of towns in such a hierarchy of settlements (often termed a hierarchy of central places), apart from that of the level one centre, depends on the size and productivity of its hinterland (the area for which it is the central place) and the portion of production which it retains; the size of the level one centre is a function of the size and productivity of the whole area in its system plus the portion of the production which it has to allocate to its supporting local élites and other functionaries at the lower levels of the hierarchy.

The situation portrayed in Figure 3.1 and described in the preceding paragraph is clearly an idealized one, for conditions are never so uniform over such a wide area as to produce the same number of surplus food units from each cell. Variations in the productivity of either or both of land and labour will result in differences between level three centres, and probably also level two centres, in the number of units available for consumption there, and thus the population that can be supported. Similarly, the élite in the level one centre may be prevailed upon to allocate a greater portion of the surplus to the colonial élite in some areas (presumably those most difficult to control and maintain within the system) in order to retain their loyalty. Together, these two factors will mean that rather than a strict hierarchical system of settlements, with all those at a particular level having the same population, the populations of centres at a certain level will vary according to the volume of surplus which they control and the proportion of that which they are allowed to retain. Thus the hierarchical organization will be represented by a modified hierarchical set of settlement sizes. (Further modification would result if some of the production was of raw materials other than food; metal mining, for example, would probably create substantial clusters of producers larger than the smallest central places of level three.)

Maintenance of such systems, particularly the loyalty of colonial élites and producers some distance from the main centre, is difficult; secession of cells is always possible, and the élite in the level one centre may have insufficient strength to oppose it, especially if there is simultaneous trouble in two or more cells. Loyalty is best obtained by taking only a small portion of the surplus production back to the 'heartland', but ensuring that in absolute terms this is large. For the latter, efficient production, a good transport system and an effective administration are all necessary. Even more necessary, however, was the change of societal organization to mercantilism.

Mercantilism and urban systems

Urban systems established and operating under a system of rank redistribution, as just described, depend for their continued existence on the ability of the élite and its various sub-élites to extract the demanded surplus production from the subservient populations. This they must do by a combination of persuasion (religion) and force (militarism), which has to be applied at all levels (sub-élites must be kept under control by the central élite, and the former, in turn, must maintain control of their local producers). Mercantilism introduced a major change to this need for absolute control, however, with a system in which producers could be persuaded that creation of a greater surplus, which could be sold and the returns used to purchase other commodities, promised immediate returns, in addition to the intangible ones offered by religious exhortations and military might. They remained subservient to the élite, but were encouraged to produce for individual as well as for élite ends.

Mercantilism operates by the creation of a merchant group which organizes the buying and selling of commodities and receives the difference between the buying and the selling price as the payment for its services. This is, in effect, part of the surplus. Under rank redistribution, all of the surplus yielded by the food producers went either to the élite or to those necessary for the élite's power. Merchants did not necessarily reduce the flow of surplus to the élite. In part, they replaced some of the former 'support staffs', and thus merely reoriented the flow of the surplus. Much more, however, they aimed to create a greater volume of surplus, by generating the conditions conducive to increased productivity; in this way they were able to support themselves, and the élite, without whose permission they could not trade, were provided with a more substantial basis for their own aggrandizement.

Initially, mercantilism involved the merchants travelling from place to place, buying up the surplus produce and selling other goods in return. The frequency of their visit to each place would depend on the volume of production there and the nature of the product: wool crops might require only annual visits, for example, whereas other animal products might support more frequent trips. At early stages of mercantile development, therefore, usually preceding the widespread adoption of a monetary system, the merchant class would be small and its impact on the urban system slight.

Increases in production, in the range of products and in the number of people involved in the system all create greater demands for mercantile services. As individual producers become more specialized in their activities, and thus more dependent on others for a variety of products necessary for daily life, so they will require to make more frequent purchases of the means of subsistence. Merchants provide for this, not by visiting the home of every producer, which would be inefficient in the use of time, but by foregathering at specified settlements on fixed dates at fixed times; these merchant gatherings are the articulation points for trade. If the area in which they are operating has no permanent settlements of any size, or if for some reason they are not allowed into such settlements, the merchants may gather, and the producers (who are both sellers and buyers, although the former may apply less frequently than the latter) visit them, at some accessible point. Both of these, and the former is the most common, are most efficient if all participants are

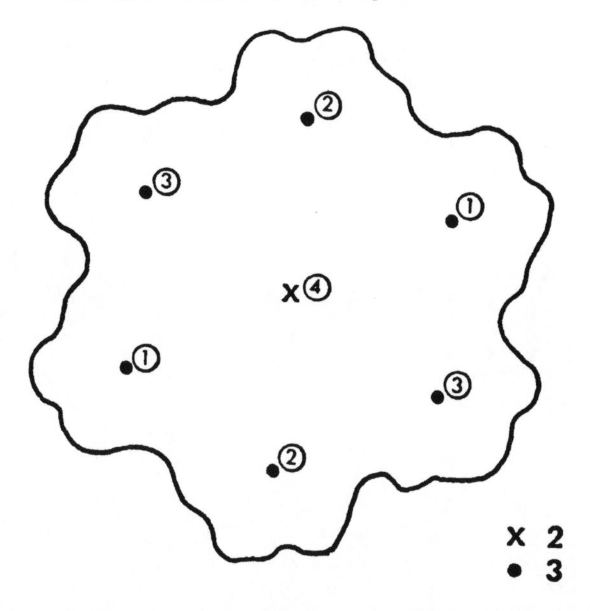

Figure 3.2 Two market circuits within one of the cells of Figure 3.1. The number alongside each settlement indicates the day of the week on which a market is held there.

aware of the meeting times and dates well in advance, and a set calendar usually develops of market days when merchants gather at certain places.

A large variety of these calendars has developed. Figure 3.2 illustrates one possibility, for a single colonial cell of Figure 3.1. The market calendar operates on a five-day week. On each of the first three days, there is a market in two of the level three centres; on the fourth day, there is a single market, in the level two centre; and there is no market activity on the fifth day. Each of the markets in the level three centres is attended by a group of merchants, who may travel out to the settlement from the level two centre or who may make a 'weekly' circuit around the four places, returning to the level two centre on the fourth day, where they meet the merchants operating the other circuit within the cell. The market in the level two centre is the most important in terms of size; by bringing together merchants who buy and sell in different parts of the colony, it enables them to trade among themselves, obtaining products to be sold in their next circuit of the lower-level centres. Further, at that major market they may also make contact with traders operating a circuit around the second-level centres, as depicted on Figure 3.3, who return to the level one centre every fourth day. These will purchase local

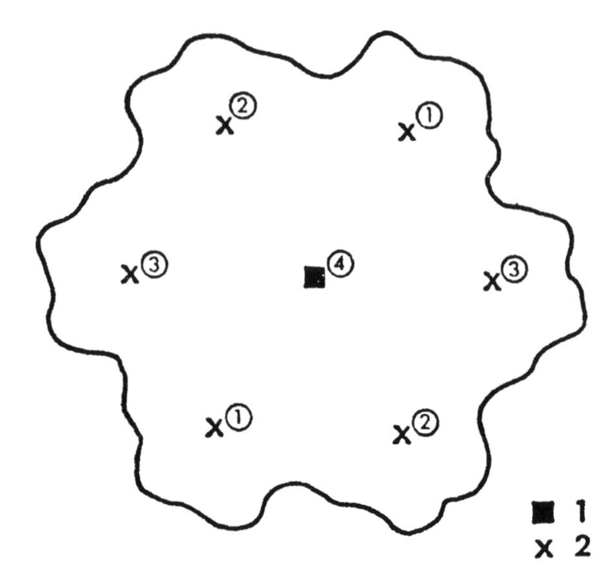

Figure 3.3 Two market circuits operating in the system of Figure 3.1, in the level one and level two centres. The number alongside each centre indicates the day of the week on which a market is held there.

products from the merchants who operate within the colonial cell, and will sell to the latter goods obtained elsewhere in the system. Thus the hierarchical system of settlements operates as an interlocking hierarchical system of markets, articulating the flow of goods between distant producers and consumers, through a series of intermediate bulking points.

Such market systems may have been inaugurated on an ad hoc basis, but as the volume of trade increases and the number of merchants operating the various circuits grows, so the calendar will become formalized and the conduct of the markets institutionalized. The latter process will involve the merchants treating with the local élite in each centre. For the latter the existence of the markets, and the volume of commerce being conducted, will be seen as a potential source of income, and traders will be charged a portion of the surplus in order to obtain the privilege to conduct their operations within the settlement governed by the élite. This system of licensing will be acceptable to the merchants if in return they are afforded protection from competition by a limitation on the number of licences to trade which are issued. Thus the local landowning and controlling élite and the merchant class derive mutual benefits from control of the trading function of the settlements.

Trade is articulated in early mercantile systems, therefore, by an interlocking series of merchant circuits around market systems. Figures 3.2 and 3.3 illustrate two interlocking circuits only, but as trade expands, and products are moved over longer distances, although perhaps only in small quantities, so the number of interlocking circuits may grow too. At the highest level, visits by merchants to each place may be rare, perhaps only annually to purchase the year's production; such a visit is likely to last several days, and may involve the merchant buying in one place the produce of several cells which has been collected in that place through a series of minor circuits such as those illustrated in the figures. At this level each merchant will be highly specialized, concentrating on the buying and selling of a single commodity only, which he may transfer over considerable distances. Where the interlocking circuits meet at the largest settlements, therefore, the number of merchants and of separate commodities being traded may be very large and the meeting continue for several days.

As mercantilism develops, so the system of settlements set up to establish élite dominance over its subjects is translated from a hierarchy of control centres into a central place hierarchy, in which the control function is only one of several. The volume of trade conducted at each place is, like its size in the rank-redistribution system, a function of the size and productivity of its hinterland, and variations in productivity levels between cells of similar size will be reflected in variations in market size. In general, the larger the settlement was under rank redistribution, the larger it is likely to be under mercantilism, so that a hierarchical pattern, modified by environmental conditions, is likely to be repeated in the merchant activities.

With the merchant activity increasing, so there will be a growth of other functions that is reflected in the settlement sizes. These will include administrators to oversee the operation of the markets, transporters who move the increasing volume of commodities, and craftsmen who manufacture the increasing range of tools needed for production and transport, as well as the consumer goods demanded by those who gain large volumes of the surplus. Some of these functionaries may move around the market circuits with the merchants; others will find sufficient custom in a single cell for them to remain at a particular centre and they, together with those who provide the market infrastructure – the administrators plus those who maintain the roads and other transport links – form the basis for the establishment of permanent trading functions.

From temporary to permanent marketing functions

As the volume of trade in a centre increases, so the permanent population resident there will grow accordingly, comprising both those who facilitate the trade and those who are parasitic, or partially so, on its proceeds. The greater the number of residents, the greater the volume of locally generated business, and this may become sufficient to support a merchant who can establish a permanent trading post there and need not move around the market circuit in order to find sufficient business. These permanent premises are the forerunners of modern shops, and the resident consumer potential may eventually be sufficient to support a constellation of such shops, providing a range of functions on a daily basis for both the residents of the settlement and the inhabitants of its hinterland. The translation from periodic marketing to permanent trading will not occur simultaneously for all functions, however, because the volume of trade will not be sufficient to support shops. Thus the former system will survive, alongside the new one. Markets will continue to be held, expanding the range of goods and services available for sale on a particular day and probably, because it will attract in producers from the local hinterland, generate the busiest trading day for the permanent establishments. (Some of the latter, notably those associated with manufacturing functions, may indeed only operate their shops on that day, and spend the rest of the week on other tasks.)

The higher the level in the hierarchy of central places occupied by a settlement, and thus the larger its population, the earlier the development of some permanent trading establishments is likely to be. Similarly, the smaller the place, the narrower the range of permanent establishments which it can support, and the greater the dependence of its residents (plus those of its contributing hinterland) on the periodic markets. Thus the hierarchy of central places will be identifiable not only by the size of the market in each place but also by the number of permanent shops there. Residents of the places which do not possess certain establishments must either await the visit of a relevant trader to their home settlement (or to their nearest central place in the case of producers in rural areas) or must travel to a larger centre. The time and cost of travel will mean that if the latter option is taken up, the almost certain choice will be of the nearest centre offering the needed establishment, and given the hierarchical organization of central places, this will mean the nearest centre of the particular level which supports such establishments. Such a set of choices for one resident of the central place system in Figure 3.1 is given in Figure 3.4. His journeys to the nearby level three centre will be

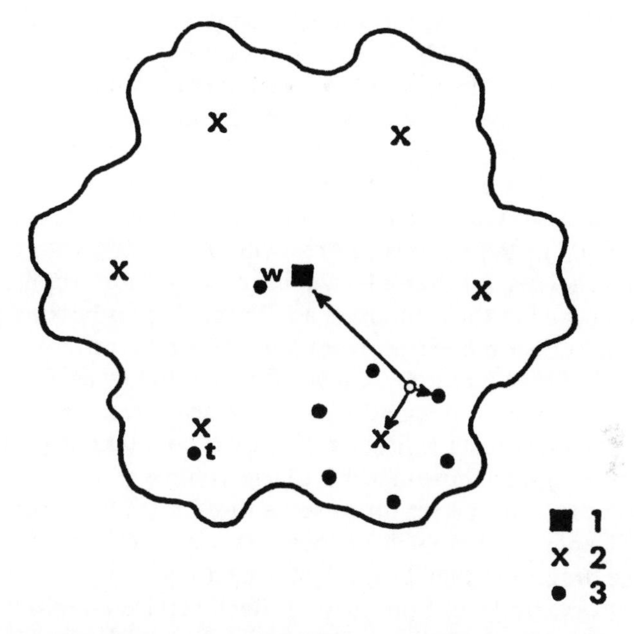

Figure 3.4 Hierarchical shopping behaviour within the settlement system of Figure 3.1, indicating the centres visited by a resident of o.

the most frequent, and will be undertaken to make regular purchases of goods, stocks of which must be replenished frequently. The trips to the larger, more distant centres will be for goods not available at a closer place; such goods are usually required less often than are those purchased at a level three centre, and so the trips will be less frequent. Thus the larger the place, the less frequently it is visited. (Note that this is relative to the customer. For a person living at point 't' in Figure 3.4 a second-level centre is his closest, and he will purchase there what others may obtain from a third-level centre; for the person living at 'w' all of his requirements can be met from the nearby level one centre.) And, as the inverse of this, the larger the centre, the greater the area from which it draws customers to particular establishments, but the less frequently they visit it: level three centres are the focus of frequent, perhaps weekly, trips by residents of their immediate hinterlands, whereas level one centres, as well as attracting a local flow, also benefit from the infrequent (perhaps annual) expeditions for rarely purchased goods undertaken by residents of distant parts of the system. (The latter may tend to congregate at the level one centre on certain dates, such as a major market or religious festival.)

As growth proceeds within the economic system and as individual producers and households find it necessary to conduct an increasing volume of trade, so the demand for permanent establishments expands and the marketing system will decline in importance; its major function will remain in the buying of the annual yields from producers while the permanent shops will cater for the weekly and monthly expenditure of the income based on those sales. The number of other types of permanent establishments in the various centres will grow along with that of the shops. Warehouses, for example, will be required for the bulk storage of products, both after their purchase and in the various bulking processes undertaken in their movement from one area and centre to another, up and down the urban hierarchy. Craft industries, too, will develop to meet the widening range of demands from producers and consumers; their workforce requirements will boost the town populations. But in many places, and especially those in which a large volume of trade is conducted, other businesses will be set up to serve those already present. As outlined in Chapter 2, these will include financial and legal services, plus specialized transport facilities, all of which encourage investment in, and growth of, trade and production. Such service establishments not only increase the population of the towns in which they are located directly, through the households of the owners and employees, but they also generate other growth impulses, for these residents themselves create demands for shops and other establishments (see below, p. 102). Business growth generates more business growth, therefore, so that the more a town has, the more it gets. And because the administration of the settlement, by the élite, is associated with this growth, and benefits from it through taxation and other charges, mercantile expansion encourages élite expansion too. Because the bigger the place the greater its growth potential, the higher levels of the urban hierarchy benefit most from such growth, widening the difference between them and the subsidiary lower-level centres in terms of population. (In part, the extent of the gap depends on the policy of the central élite: the greater the control which they exert, and the less independence they allow the secondary élites in the 'colonial' centres, the more the growth impulses are concentrated in the one centre.)

Mercantilism is reflected in the urban landscape, therefore, by a strengthening of the hierarchical system of settlements. Such a hierarchy may develop several levels. At the highest is the dominant centre, in which are found all of the establishment types also present in lower-level centres (often termed those of a lower order), plus many others which are peculiar to that place. Each succeeding level below comprises places with smaller hinterlands and able to offer only a narrower range of establishments. At the lowest level of all are settlements with but a rudimentary central place content, able to provide only the goods and services most frequently bought by all. No such system has developed on a uniform plain of invarying productivity, as suggested in Figures 3.1 to 3.4, and so the hierarchical pattern is modified to meet local circumstances. The smallest centre in a productive area may be able to support several more establishments than one where the land is much less fertile, therefore, but although the resulting pattern is not a clear-cut hierarchy of settlements, in terms of their populations and functions, the organizing principles remain the same.

The urbanization process

The previous sections have described the growth of urban places within a settlement system, but have not indicated the sources of that growth in terms of population origins. In such situations, urban population growth is dependent on increased productivity in rural areas, both to provide the necessary surplus to support the non-food-producers and to provide the urban residents themselves (or at least many of them; others will be the children of the immigrants).

The foundation of urban growth is increased productivity in rural areas, and this can only be achieved by improving the capabilities of the food-producing workforce. Better tools and better raw materials (improved seed strains and breeding stock, fertilized soil, etc.) are needed for this but, as pointed out earlier, there is a law of diminishing returns for the productivity of each unit of labour. To circumvent this, the labour force must be reduced, whilst those who are retained are provided with the ability to produce even more. This not only increases the total production, enhancing the ability to support non-food-producers, but also increases the potential returns to those who invest in agriculture (if more can be produced by fewer, the potential surplus is greater and the attraction to investment substantial). Those who leave the agricultural labour force, perhaps having been forced out by the élite landowners who wish to invest in higher-yielding methods involving fewer workers, can move to the towns, where there are employment opportunities both in handling the increased volume of trade and in manufacturing the tools required by the remaining agricultural workers (plus providing the products demanded by the enriched recipients of the surplus). These ex-agricultural workers will now be consuming part of the food surplus, rather than producing it, but because the total production is increased by their transfer out of the agricultural sector, their new activities make it possible for an even greater surplus to be produced.

The process just described is commonly known as urbanization, which involves the movement of people from countryside to town and a consequent increase in the relative proportion of a system's population resident in its urban places. There should be an equilibrium in the population distribution at any one time, reflecting a balance between the declining demand for agricultural labour (consequent on the increased productivity of the remaining, better-tooled workers) and the increased requirements for workers in urban occupations. Thus rural depopulation fuels urbanization in a symbiotic relationship: the rapider the rate of economic growth in the system, and thus of expansion of the food surplus from rural areas, the faster should be the rate of urbanization.

This equilibrium may not always hold, however, with resulting problems for at least some of the population. Thus the rural areas may not be releasing workers rapidly enough to meet the demands in urban areas, so that urban employers may be forced to pay higher wages to their workers. Without greater productivity in return, this will make products relatively expensive and reduce their sales in rural areas, with probable declines in demand. More importantly, and more likely, the agricultural sector, in the landowners' search for greater productivity and profits, may be releasing more workers than the urban areas can absorb. With no work, or likelihood of it, available in the rural areas, these unemployed persons, many of whom may have been dispossessed of land holdings by the profit-seekers, will probably drift to the towns where there exists the possibility of occasional work and access to a small part of the surplus. The existence of such a pool of job-seekers enables the urban employers to hold down wages close to subsistence level, and so ensure large portions of the surplus (profits) for themselves.

System expansion: Mercantilism and colonialism

Growth in systems such as that just described can be slowed by a shortage of any of the resources used in production, notably of land and labour. A shortfall in the number of workers can be countered by investment in productivity increases and by bringing into the pool of potential employees those formerly outside it, such as children and women; if both are insufficient to meet demands, then extra labour must be sought from beyond the system, involving some immigration (either voluntary or, if necessary, forced). Such labour shortages are likely to be relatively short-lived, however, and relatively easily countered. Much more important are the shortages of land-based resources which, with prevailing technology, cannot be countered beyond the level set by the law of diminishing returns. The only solution to the demands for more food and other raw materials then involves expansion of the system. This expansion, where it involves permanent involvement in land not previously occupied or traded with, is the mercantile form of colonialism, the process introduced earlier for rank-redistribution societies.

Local colonial expansion

In the search for more food and other resources, merchants invest in trading expeditions which venture outside the area controlled by the élite of the system in which they live and trade. Such expeditions require investment; if they are successful they will provide a financial return whose size may be sufficient to encourage further investment in similar ventures.

Many such ventures may be into the territories of adjacent systems, and generate trade which is beneficial to all. The two sets of merchants involved may be bargaining as equals, however, which will not allow one group, and their backers, to make large profits. More attractive, therefore, is trade in which those seeking resources are the stronger, and can extract the surplus from their new contacts at a relatively low price. For this, rather than for the individual expeditions to trade with equals, the merchants require military support, in order to establish their hegemony over those being drawn into their trading sphere, and such expeditions require the involvement of the state in the trading and colonial ventures. After military conquest, administrators will be needed to operate alongside the continued armed presence, to ensure the merchant success, and religious support may also be insisted upon. In this way the élite of the 'home society', through the state mechanism which they operate as landowners, become involved in guaranteeing the success of mercantile ventures; where the potential for mercantile success is considerable, this support will be continuous, and lead to permanent settlement of the colonized land. The state organization will be expanded, therefore, and this must be paid for out of the extra surplus, through taxes and other charges on the merchant venturers and those who benefit from their actions. (The taxation may extend across the entire 'home' population, many of whose benefits from the ventures are few and unseen.)

This militarily supported expansion of trading activities may be into

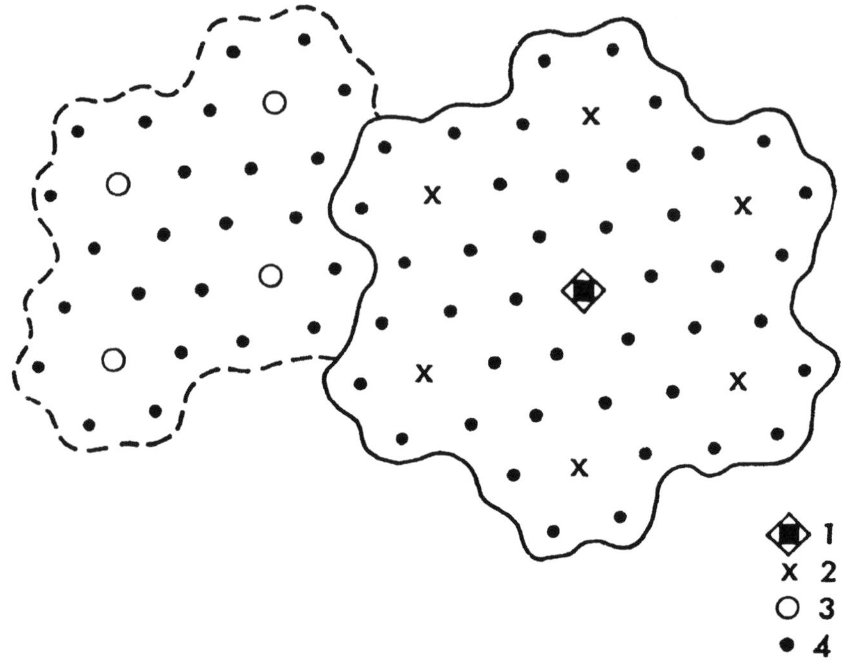

Figure 3.5 The hierarchical system of settlements where a colony of four cells has been added to the original territory.

adjacent lands, in which case settlement systems will probably be established similar to those shown in Figure 3.1. Extra colonial cells will be settled (Figure 3.5), and the surplus derived from them will be channelled back to the source of the investment – in all probability the level one centre in the original system, whose growth will be boosted accordingly. Such an eventuality is unlikely, however. Much more probable is that the adjacent lands are already occupied and organized by an urban network comparable to that of the colonizing power. The role of the colonizers will then be to re-articulate the flow of the surplus in the conquered areas, diverting at least a portion of it away from the local élite and back to the investors in the homeland. (The local élite may be allowed to retain a portion in order to secure their loyalty to the colonizers – and thus reduce the costs of military occupation. If the colonization generates growth of productivity, this portion may, in absolute terms, represent an increased yield to the local élite.)

The effect of this colonization on the urban hierarchy will be to create further steps in it. The towns in the colonized area will have less of the growth potential than their counterparts within the original system. As pointed out in the previous section, growth begets growth, so that an increase in the volume of trade in a town generates work for others there, as the surplus produced is invested in other activities. If some of that surplus is not retained locally, however, but is repatriated to the home towns of the colonists and those who invested in them, then some of the growth impulses are transferred away from their origins and into the other centres. Figure 3.5 provides a possible scenario for this, which involves the expansion of the initial system (enclosed in the solid line), with its three-level hierarchy, into a four-cell colony to the west. Each of those cells is controlled by seven towns. Those at the lowest level will probably be the equivalents, or nearly so, of their counterparts in the colonial heartland, for only a small portion of the surplus is retained to generate growth in such places: thus they are all placed at the fourth level of the expanded hierarchy, whether or not they are in the colony. In the heartland the main centre of each cell retained 40 per cent of the surplus, sufficient to generate considerable local growth. Their equivalents in the colony are only allowed to retain 25 per cent, however, with the other 15 per cent being repatriated by the merchant venturers. This both stunts the growth of the colony towns, placing them at a lower level of the hierarchy than their counterparts, and generates further growth in the capital, whose size relative to that of the second-level centres is enhanced accordingly.

Other changes to the settlement pattern are possible, but are not indicated on Figure 3.5. For example, the links between the level one centre and the colony will probably operate through the level two centres lying between it and the conquered area; these centres may gain extra growth impulses as a consequence of the traffic passing through, and outpace the other four level two centres in rate of growth. Further, rather than incorporate the colonized cells into the original system as an integral part of the 'national territory', the colony may be given some separate identity, with its own rulers operating from one of the four level three centres, thereby probably raising it to the equivalent of a level two in population. The variations are many; the results are that the expanded system almost certainly will not have a strictly demarcated hierarchy of urban settlements, even though the hierarchical principles of organization remain. The only certain consequence is the increased wealth of the residents of the level one centre, from which the colonization is controlled, and its consequent increased growth relative to that of other centres in the original system. Its merchants, bankers, craftsmen and others will all benefit from the increased volume of trade generated by the colonization, and any of these will seek to ensure, by restrictions on activities there, that comparable functions do not develop in the conquered areas. (If the colonized area is eventually assimilated into the national territory, however, these restrictions will have been channelled elsewhere so that entrepreneurs in the stunted towns will compete with their longer-established rivals at great disadvantage.)

Distant colonialism

The previous example of local colonialism and the conquest, followed by the assimilation, of adjacent areas represents many aspects of the changing urban settlement system in certain parts of the world (notably Europe). But the potential for such colonialism was limited, and much more common, especially in the later stages of the process from mercantilism into industrialism, has been colonial expansion into more distant lands, previously occupied at much lower levels of economic development. In most cases this expansion has been trans-oceanic rather than overland, because of the greater ease of movement over sea with relatively limited technology, especially movement of bulky cargoes.

Two types of colony have been established in such situations. In the first, the colonists encounter a relatively dense distribution of population, in societies operating at either the reciprocal or the rank-redistribution levels. After conquest these societies cannot be eliminated, and so a settlement system must be established within them, perhaps involving the local élite to a minimal extent. The colonizers are unlikely to become permanent residents of the new areas, however; instead, the merchants, administrators and the military will comprise employees sent on tours of duty to the colonies, and no permanent ties to the new land will be forged. The urban system will reflect this. The main town will be a port, through which all of the surplus flows. Virtually all of the profits from this will be retained in the colonial heartland and very little growth will be generated; much of the expenditure of their salaries by the temporary residents will be on imported products. If the colony is large, other settlements will be necessary to transfer the surplus to the port. These will be populated by as few merchants, administrators and military as possible; the amount of money circulating will stimulate virtually no growth of functions. Thus the towns will be almost exclusively parasitic way-stations or depots in a process of wealth-creation for the élite of the homeland and generating virtually no wealth for the permanent residents of the colony. (Indeed, the latter societies may be severely dislocated.) In some cases the colonists established enclaves (plantations) in which they produced, using labour 'imported' from the adjacent area, the materials not obtainable in the homeland; their contact with the surrounding society was very small in such situations, and brought extremely few benefits to it.

The other type of colonization involves encounter with a virtually unoccupied 'new land', in which the colonizing power has to establish a population to extract the needed surplus. This population comprises immigrants from the homeland, who are either granted or leased land (sometimes the grants are sales at low prices) on which to establish productive farms serving their compatriots 'back home'. In such cases the military presence need be less pronounced than in the other type of colony. Administration will still be necessary, however, and although much of the profit from the colonial investment will be channelled back to the homeland, a certain amount of growth will be allowed in the colony, to provide the needs of the settlers and to support them at a much higher level of subsistence than that of the natives of the other type of colony. Because many of these settlers may become permanent residents, some growth impulses will be allowed, but dependence on the 'home' economy will still be great, in large part very often because of the relatively small size of the colonial markets.

The settlement patterns in such settler colonies will be very similar to those in the other type of colony. The dominant place will be the port, the point of contact between colonial producer and homeland consumer. Indeed, the port will be the gateway to the colony. Inland from it will be a series of central places, gateways to the various parts of the colony, collecting their produce for transfer to the port and distributing imports from the homeland. The size of these smaller gateways, like that of central places everywhere, will be a function of their hinterland size and productivity. Many of these settler colonies have been occupied at relatively low population densities, because of both the nature of the local environment and the dependence on the latest agricultural technology which has not encouraged the development of a dense settlement network. As a consequence, compared to the port gateways these internal central places will probably be small. The former, containing the port and administrative functions, plus those others generated by local demand and entrepreneurial activity, will be primate in their colonies, with populations very much larger than those of any other settlements.

Since most colonies are coastal, the usual development of urban systems has a form similar to that portrayed in Figure 3.6 for a series of colonies established along the eastern coast of a landmass. Each of the five colonies shown has a single town, which is a port connected to the homeland by shipping routes. The only other element of the settlement

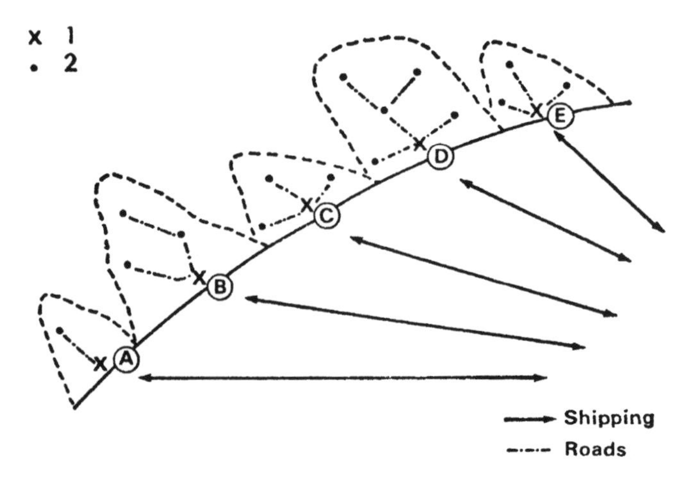

Figure 3.6 The settlement pattern in five colonies arranged along a coastline.

pattern is the small internal gateway village; the number of these varies from colony to colony according to its size and fertility. A similar pattern is likely in the colonies of the first type, except that the small village gateways will be surrounded by relatively dense areas of native occupancy, whose production is articulated at those points.

The establishment and development of this colonial urban system

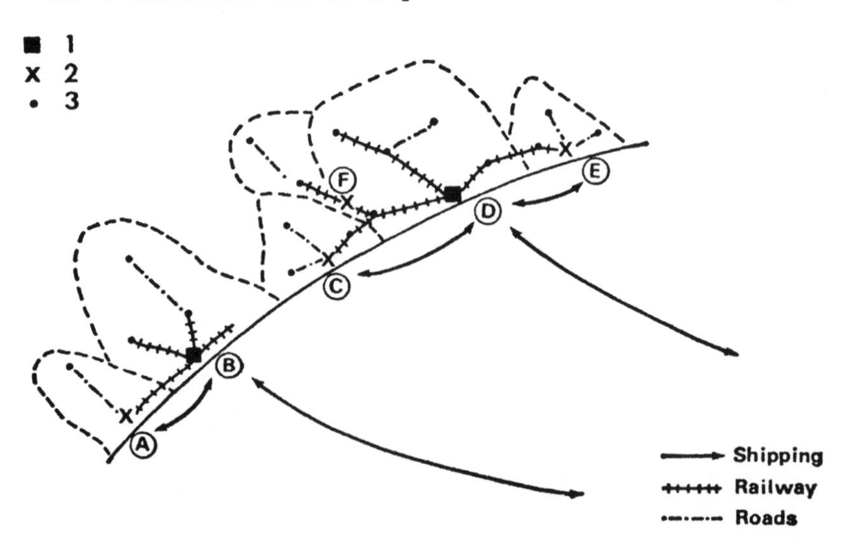

Figure 3.7 Later development of the settlement pattern of Figure 3.6.

follows a typical sequence which is outlined in Figure 3.7. This involves consolidation of the formerly separate colonies, which may have been operated by independent companies, into a single territory with its own state apparatus, possibly subservient to that of the colonizing power. Much of this consolidation is based on economic realism. Thus for shipping services, as the size of vessel is increased to improve efficiency while most colonies offer only small cargoes which it is inefficient for the large, expensive-to-run vessel to visit the port for (assuming that it can dock there), certain ports become the foci of international connections, and these are almost certainly those with the largest and most productive hinterlands (B and D in Figures 3.6 and 3.7); cargoes from the others are sent there by coastal shipping, for bulking before collection by the international carriers, and imports are distributed to the smaller places on the return voyages. Thus a three-level hierarchy is established, as indicated in Figure 3.7, with the two international port gateways establishing a superiority over the other towns.

This hegemony of a few of the port gateways only is furthered by the development of inland transport routes. Railways are major investments, and so are constructed where trade potential is greatest. This favours ports B and D, especially D, which is able to tap not only its own hinterland but also that of its two neighbours, thus channelling even more of the trade through its port and depressing the economic potential of its captured rivals. All of this trade passing through the port generates other activity too, although much of the profit from the colonial ventures may be retained in the homeland. Thus, as places B and D grow through their trading activities, they generate demands for local industries and stimulate the same 'growth creates growth' syndrome that characterizes all urban systems, although their partial parasitic nature ensures that some of the growth takes place elsewhere. In time, political independence will generate even more growth in the capital of the new country – probably, though not necessarily, D -- and the large places at level one in the hierarchy develop as the main urban centres. The only major change, as suggested by Figure 3.7, occurs when a new inland area is colonized, as between the hinterlands of C and D. This new colony is organized through an 'inland port gateway', which grows into town F at the second level of the colonial urban hierarchy.

The development of an urban system in colonial areas is accompanied by consequential changes in the urban system of the 'homeland'. Two places in particular are affected in the latter area. The first is the main urban centre, the dominant central place and, in all probability, the political as well as the economic capital. Colonial enterprises require a great deal of finance, which must be aggregated from a variety of sources, who are usually the bankers of the main city plus the landlords on whom the merchants depend. Thus the organization of colonization creates activity in the main city, which generates other growth; successful colonialism returns to that city profits which are the source of yet more growth. Thus the size of the political and economic capital of the colonizing power increases in proportion to colonial success, so that its growth outpaces that of its subsidiary central places which are dependent on local production only.

If the capital city is also the colonial homeland's major port, no other place may benefit directly and in large amounts from colonial activity. If it is not, then the port settlements, some of which may have been small places in the central place system that preceded the expansion of overseas trade, will benefit too. Immediate beneficiaries will be the traders, those who service and finance them and those who provide the shipping. Their returns generate income for other activities in the port, and initiate the growth syndrome there, stimulating rapider growth than is experienced elsewhere in the system, except in the capital.

The symbiosis between these two settlement systems is illustrated by the ideal type-sequence outlined by Jay Vance (Figure 3.8). In the initial stages of mercantile exploration no permanent settlement is established in order to obtain the required products (fish, timber and furs). Then the colony is settled by agriculturalists; the export of their products moves through local articulation points to the colonial port, and thence to the port in the homeland, which grows in size and status relative to its inland competitors. As settlement of the colony expands further inland, so both of the ports increase in size, railways replace rivers as the main traffic arteries within the colony, and internal gateways develop to articulate the trade of areas some distance from the port, while in the homeland places near to the original port benefit from the imports and a new outport is built to handle the larger volume of trade and the bigger vessels. As with the other diagrams in this chapter, this is an idealized representation (or model) rather than the detailed portrayal of an actual example. It contains the main elements of the urban patterns which have developed and highlights the salient features (notably the dominance of the port cities in the colonies), all of which are modified, to a greater or lesser extent, in particular situations.

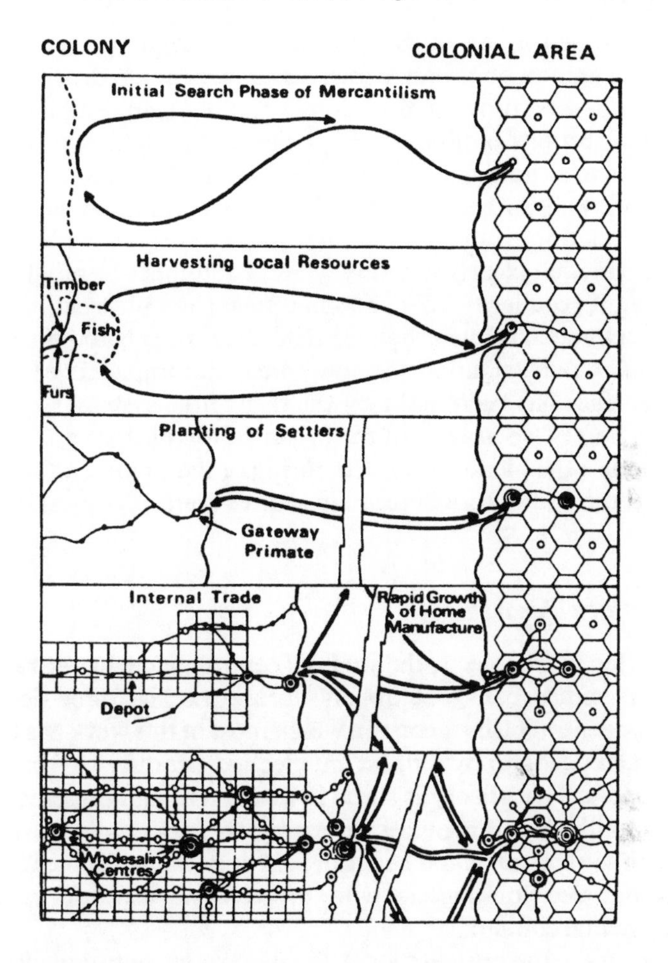

Figure 3.8 Colonialism and urban settlement patterns (after Vance, 1970).

From model to 'real world'

In a book of this length, on a topic as large as urban geography, it is possible neither to describe every urban pattern in the world and account for its development, nor to quote each as an example of a particular type of process or pattern. By providing a general outline – the model presented here of the evolution of an urban system through reciprocal, rank-redistribution and mercantile forms of socio-economic organization – a framework has been structured within which individual situations may be studied. Thus, for example, the section on urban origins provides a base for discussing early urbanization in Mesopotamia and in the Nile Valley; the section on urban hierarchies under rank redistribution forms a paradigm for studying the Greek and Roman Empires; and the mercantile sections outline the processes and patterns which should be identifiable in Britain and its former colonies. Not all of these examples can be followed up here; instead the focus is on a few features of the model in relation to major examples. No attempt is made to verify or falsify the model; clearly it is very general and must be modified to take into account peculiarities of economic, social and physical environment.

Although much of the urban fabric of Europe is based on the inheritance of rank-redistribution societies, the major impact over most of the earth's surface has been made by the mercantile system, so that will be the focus here. Two aspects of the organization of that system stand out in the model: the hierarchical structuring of the urban pattern at a local level, and the dominance of primate centres (gateway cities) in colonized areas.

Central place hierarchies

As indicated in Chapter 1, the study of central places has attracted much attention from urban geographers during the last three decades, and indeed the central place theory on which most of this work has been based was termed 'geography's finest intellectual product' by one of those investigators. That theory is static in its content and presentation, however, and although it provides an extremely logical picture of what the pattern of settlement would be on an isotropic plain occupied by a society of rational decision-makers who brooked no deviations, it has no developmental content.

According to the present model, the pattern of central places in a fully settled area developed out of a system of administrative/religious/military control of separate cells, each, except the largest, being ruled by an élite who owed allegiance to the élites controlling the larger cells which encapsulated their own. The mercantile systems perpetuated this structure by establishing a marketing system which used the control centres as locales for buying and selling: the larger the area controlled, the greater the volume of trade and, as this volume increased, the larger the permanent infrastructure provided at the centre to articulate the movement of goods from one place to another.

If the area were of uniform productivity and was settled as suggested above, the resulting settlement pattern would indeed display a series of strict hierarchical levels, in terms of population size as well as organiza-

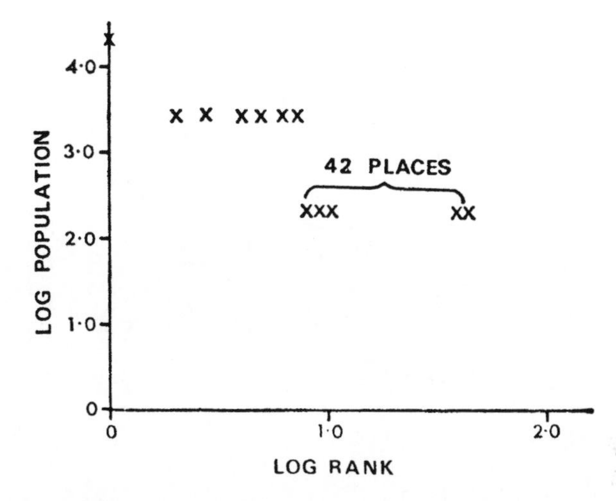

Figure 3.9 Population *v.* rank in the settlement system of Figure 3.1; not all forty-two level three centres are shown.

tion. A graph of populations against rank position in the system would indicate these steps, as in Figure 3.9, in which the axes have logarithmic scales (see above, p. 16). The populations of the centres shown are those derived earlier for the central place system of Figure 3.1 (except that very minor variations are allowed so that, for example, the six level two centres differ in their rank positions from 2 to 7, and the forty-two level three centres do not tie either). Of course, no such pattern can be expected in a far from uniform world. If the seven cells of Figure 3.1 were differentiated by productivity into two above average, three average and two below average, and the sizes of their settlements varied accordingly, the result would be a population/rank graph like that of Figure 3.10. Here the steplike separation of the levels is much less clear than in Figure 3.9 and the arrangement of centres is somewhat similar to that shown in the rank-size rule diagram of Figure 1.1; introduction of further productivity variations into the system of Figure 3.1 could produce an urban pattern very like that of the rank-size rule.

Population/rank graphs like that of Figure 3.9 are only rarely produced in studies of settlement systems, and those which have been almost all apply to small areas with few urban places. The larger the area, the greater the probability of environmental diversity and the greater the likelihood of a graph more closely approximating the form of Figure 3.10. This may suggest that the axioms on which Figures 3.1 and 3.9 were built are irrelevant to the study of urban, systems and that other pro-

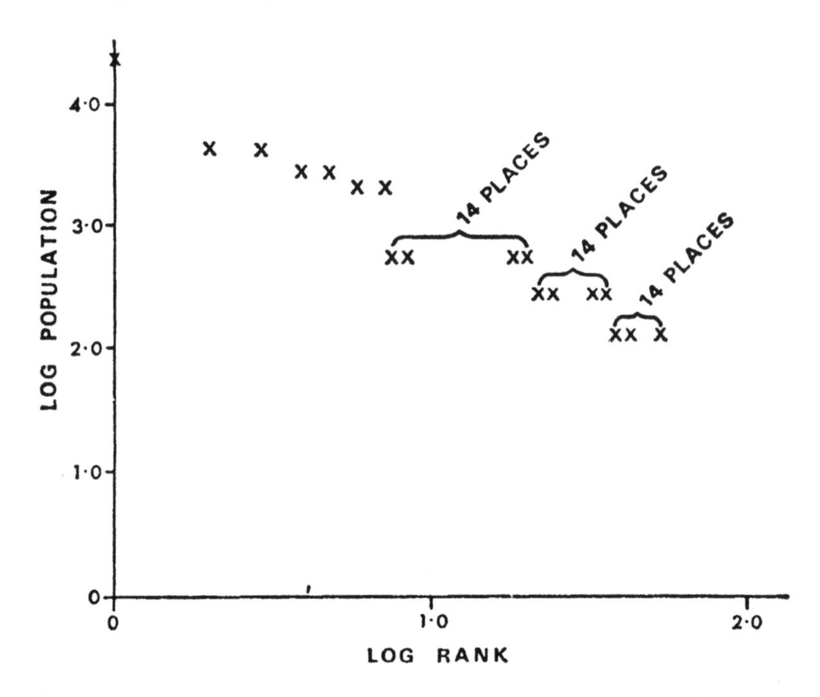

Figure 3.10 Population *v.* rank in the central place system of Figure 3.1, with inter-cell variation in productivity; not all forty-two level three centres are shown.

cesses should be proposed for the evolution of settlement patterns. An alternative view would suggest that a graph such as Figure 3.10 does not deny the relevance of the principles underlying the model presented here, but rather indicates that in a complex and varied world (in which, incidentally, industrialism has been imposed on many of the mercantile frameworks, as indicated in the next chapter) the resulting pattern is far from simple. Indeed, a perfect fit to the rank-size rule of Figure 1.1 could be interpreted as the outcome of central place principles as outlined here.

How, then, can the model be compared against reality? The main way is to study the processes underlying settlement evolution. If the model is correct, it should compare favourably with what transpired in the long-settled lands of western Europe which passed through the feudal stages of rank redistribution and into full mercantilism. In England, for example, and in much that is now modern France and Germany, the allocation of land by lords to vassals created a hierarchy of spatial organizational cells into which the mercantile system intruded. At the lowest level the lands of individual households (such as those of the lords of the manor) were small, and in only a few cases were they able to support settlements large enough to be called towns and to attract itinerant merchants. Nor were they all of the same size and supported by land of uniform fertility, farmed to the same degree of efficiency. Nevertheless, many of them, particularly those with nucleated rural settlements, contained villages which, as mercantilism grew, were of sufficient size to support a few shops and other trading establishments, thereby occupying the lowest levels of the central place hierarchy. And above them, associated with the landholdings of the superior lords, developed a system of towns, initially associated with sequences of periodic markets and fairs and then with permanent trading. Analysis of these – in East Anglia and elsewhere – indicates that they did indeed operate as central place hierarchies, in the circuits traversed by the merchants and in the flows of goods. More recently, examination of such periodic marketing systems in areas still dominated by mercantilism (as in parts of West Africa) is providing even more evidence of the general validity of the model. Other evidence deriving from present-day behaviour shows that consumers, especially rural residents in many parts of the world, use settlement systems in the way suggested by Figure 3.4. The principles underlying central place modelling are valid, even if the patterns observed are obscured by the influence of other variables.

The major feudal areas which developed mercantile urban systems expanded into adjacent areas in the fashion outlined by the colonialism model of this chapter. The inferior role of the towns in the colonized areas, because of their parasitic nature and the channelling of so much of the surplus back to the 'homeland', is identifiable still in settlement patterns. A prime example is France and the hegemony of Paris, which was founded on its dominance of a growing system as control was imposed on the areas beyond the fertile Île-de-France.

Central place principles were typical of much of feudal Europe, therefore, and were introduced from there to the colonial areas which Europeans colonized under mercantilism. In much of Europe the patterns they created were later obscured by the superimposition of an industrial urban system, but the antecedents remain, particularly in the rural areas where shopping behaviour still tends to be hierarchically organized. Elsewhere, from China to Chile and from New Zealand to Newfoundland, mercantilism has not been so eroded by industrialism, and the central place pattern remains dominant, within a system controlled from major central places, the gateway ports.

Gateway primate cities

Whereas in much of western Europe mercantile urban systems were superimposed upon rank-redistribution settlement patterns, in most of the rest of the world there was no well-developed urban palimpsest and, as a consequence, the mercantile impact has been the major influence on the current system morphology. In these lands the central place principles were relevant, but the funnelling of the export products into the ports, and thence to the colonial homelands, focused attention on the gateway cities and led to their dominance of the extra-European urban scene.

The sequence of urban distributions suggested by Figures 3.6 and 3.7 can be identified in much of the Southern Hemisphere, as well as over much of Asia and northern Africa. In both New Zealand and Australia, for example, settlement was initially in a series of independent colonies, most of them small and limited to the coast by difficulties of traversing the inland terrain; in almost every case, the colony was dominated by a single port town. (The exceptions occurred where difficulties of terrain limited urban growth at the port, and so the main urban centre was developed a little way inland, adjacent to the more fertile areas.) Initially, there was very little contact between these colonies, and each had stronger links with Britain (and with Australia in the case of New Zealand) than with its neighbours. With time, however, the pattern illustrated by Figure 3.7 developed, as shipping services were concentrated on the ports serving the more productive hinterlands and investments in railways helped those ports to capture trade from their smaller competitors. (The pattern of railway investment was often strongly influenced by political considerations, with the main administrative centres being favoured with the early connections that gave them a headstart in the inter-port competition for trade.) Thus New Zealand's trading links with the outside world were soon focused on the four centres of Dunedin (serving the south of the South Island plus the goldfields of Otago), Christchurch (serving the fertile Canterbury Plains), Wellington (the national capital) and Auckland (centre for the most productive areas, in the northern North Island). These four retained a primate position within their respective spheres of influence; in each of the sub-regions which they served, a relatively small centre exercised similar primate dominance in a more limited way. And then, since the Second World War, the economy has become increasingly focused on Auckland alone, which is moving towards a primate position.

Australia is much bigger in area and in population than New Zealand, and this is probably a major reason why it has not developed a single primate centre. Instead, each of the states has its own primate centre

(Perth in Western Australia; Adelaide in South Australia; Melbourne in Victoria; Sydney in New South Wales; Brisbane in Queensland; and Hobart in Tasmania); the lowest levels of primacy are in those states (Queensland and Tasmania) whose capital cities are eccentrically located relative to the bulk of the most productive farmlands so that, for example, in Queensland there are sizeable urban centres along the coast at Rockhampton, Cairns, Townsville and Mackay. Another reason for the lack of a national centre is the federal political structure; this was only introduced in 1901, before which the colonies were independent of each other and developed their own urban and communications systems (the latter including the variety of railway gauges which failed to join at the state boundaries).

In both Australia and New Zealand, therefore, the urban patterns are the product of the mercantile era, and are dominated by primate gateway cities. In neither case, however, does this produce a national settlement system dominated by a single primate city, unlike the situation in, say, Uruguay and many West African countries, most of which are small and are focused on a single city. The comparison should not be between Uruguay and Australia, however, but between Uruguay and, for example, South Australia, in which case the similarity would stand out: in both, the primate city, which is the organizing nexus for the hinterland, houses about two thirds of the total population.

The existence of primate gateway cities has not been recognized in several studies of urban systems: neither Australia nor New Zealand is characterized by primacy according to most such analyses, for example. This is because the modern nation-state is used as the unit of analysis but, as indicated above by the Australian example, this contemporary political territory may have been non-existent at the time when the urban patterns developed. Neither India nor China has a primate urban pattern at the present time, but both were colonized from Europe through a series of gateways, each of which remains primate within the territory which it has always served. (In India the main primates are Bombay, Calcutta and Madras.) Thus the fact that primacy can only be identified at present in some countries, since national data are employed, does not mean that gateway primate cities were not, and are not now, the major elements of the settlement pattern. In a few cases they are not, often because of the impact of industrialism (a major example of this is the position of Johannesburg in the South African urban system), but through much of the world careful analysis shows that primacy is a very general, almost universal, characteristic.

Colonialism (whether or not accompanied by imperialism) has also led to primate status for certain other cities, notably the capitals of several Western European countries which dominate their urban systems more than would be anticipated from the operation of central place principles. Major examples of this phenomenon are London and Paris, both of which have in the past been the centres of extensive overseas empires (and in economic terms they still retain a good deal of their past control). Their current size reflects a channelling of surplus from much larger areas than their present respective nation-states (some from nearby: the closest primate colonial gateway city to London is Dublin). Cities which have lost part of the territory which they controlled also display a primate pattern which is not related to gateway status: Vienna, now the capital of Austria but formerly the focus of the Austro-Hungarian Empire, is a clear example.

Inland gateways

One former colonial area which has no apparent primate urban system is the United States; indeed, this country is frequently quoted as the one that best fits the rank-size rule which, according to the earlier discussion, may indicate the importance of central place principles. Closer analysis, however, indicates that there, too, gateway primate cities have dominated, and still dominate, the urban pattern.

Figure 3.7 introduced the concept of an 'inland port gateway', which describes the colonization of an area some distance inland from a major port gateway through a single major town on its edge. For a large continental land mass whose inland areas are relatively fertile (unlike those of Australia), several such inland gateway ports may develop. Figure 3.11 suggests how. At the first stage shown (A), both of the coasts have been settled and two series of centres, five on the east coast and three on the west, have attained major status, as in Figure 3.7 and described earlier for New Zealand. Each centre has rail links to its hinterland, which will contain a system of small central places functioning as articulation points structuring flows to and from the port gateways.

At the second stage illustrated (Figure 3.11 B), settlement has been extended westwards from the east coast hinterlands, mainly in the north-central portions, and a new series of towns, marked by 'X', has been established. This series comprises the first generation of inland gateways. Each is primate within the region which it serves, and from which it draws surplus along the expanding rail network, but each is also inferior, in status and in size, to the major port with which it is linked. The latter has provided much of the capital to finance the expansion of settlement

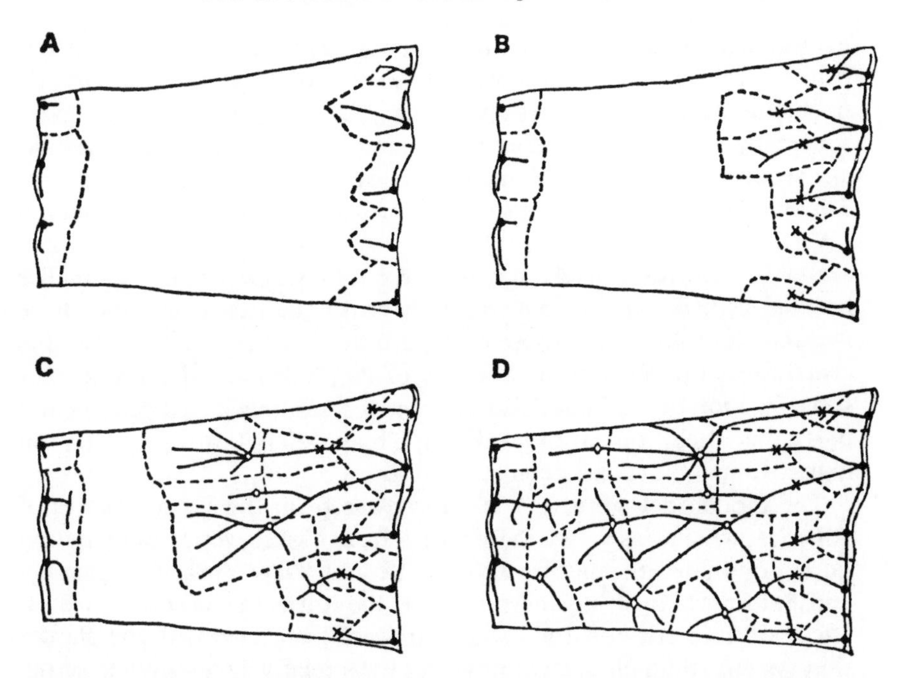

Figure 3.11 Four stages in the development of a mercantile urban system including inland port gateways.

and the necessary infrastructure, and some of the profit flows back to it, to generate further growth there. As will be seen, development has been most extensive in the new areas served by the port gateway with the largest hinterland at stage A (the second from the north) and the most extensive rail network serving it; its success at one stage is the basis for further success.

At the third and fourth stages shown (C and D), settlement has proceeded further inland, mainly from the east, producing two further generations of inland port gateways, marked by 'O' and by '◇'. Some have clearly been able to obtain a greater volume of investment in railway extensions than have others; this is reflected in the extent of their hinterlands, their size and the size of the original port gateways with which they are linked. Settlement from the west coast, on the other hand, has been relatively limited, and indeed in the north of the country the east coast ports have established dominance across almost the full breadth of the country.

As this process continues, each new generation of gateway cities, although primate over the area for which it is the focus, comprises settlements smaller than those for which they act as feeders (i.e., those established in previous stages and which have already entered the apparently unending sequence of growth generating further growth). Nevertheless some may outgrow those of earlier generations but not in the same linkage sequence, as a result of hinterland size and productivity variations. Thus the two inland gateways marked 'X' and two of the three marked 'O' (excluding the central one) linked to the major city on the eastern seaboard may, respectively, have outgrown by stage D both the smaller original centres on the southern portion of the east coast and those marked 'X' which serve those centres. Indeed, if the size of each of the centres was proportional to the size of its immediate hinterland, plus some fraction of the hinterland of other centres linked to it (and inland of it), then their population/rank graph may well indicate a distribution akin to that of the rank-size rule.

The model in Figure 3.11 is clearly based on the United States of America, which was very largely settled from east to west. Four gateway ports dominated the eastern littoral (Boston, New York, Philadelphia and Baltimore) in its northern section: the towns in the south, such as Richmond, Charleston and Savannah, were associated with the plantation system and their growth impulses were relatively few. When settlement expanded across the Appalachians, the first generation of major inland port gateways was established, and it was particularly those linked with the northern ports (such as Cincinnati, Chicago and St Louis) which grew to large size and attracted the investment in extensive feeder rail networks. As the outer edge of settlement moved further westwards still, so a new generation of gateways was founded, as at Kansas City, Dallas-Fort Worth and Minneapolis-St Paul, and a final generation (places like Denver and Salt Lake City) completed the settlement process. The main links were back to New York, which was also a major source of capital, and it, along with Chicago and, at a much later stage, Los Angeles, became the major centre of the whole country.

In some of the regions served by the first generations of inland port gateways intensive development of their hinterlands led to the growth of important subsidiary centres which drew some of the expansionary impulses away from the initial node and stunted its primacy somewhat. Cincinnati suffered from this, and with the growth of such cities as Indianapolis, Dayton, Columbus and Louisville (some of which was the result of the later industrialism period) it is no longer a primate centre. A similar situation developed on the Canadian Prairies, formerly dominated by Winnipeg. But further west in the United States, where the land could not support intensive agriculture and large-scale central place development, primacy remains: Minneapolis-St Paul remains dominant on the northern High Plains, for example, and the next largest centre in its hinterland has less than one tenth of its population.

It can be argued, then, that the primate urban systems identified in much of Africa, South America, Australia and Asia represent the impress of colonialism and its typical urban pattern where settlement has, to a large extent, and mainly for environmental reasons, been constrained to the littoral. In the United States (and also in Asian U S S R), expansion of settlement at considerable densities over an entire land mass has reproduced the gateway pattern several times over and produced (in both of the countries named) a series of regions of various sizes whose towns are ordered into a pattern which is represented by the rank-size rule. Thus the rank-size rule appears to be the end-state of two separate processes, one the creation of central place hierarchies in adjacent areas of varying productivity and settlement density and the other the completion of the settlement of a continental area through gateway cities.

Pre-industrial urbanization

It is a common, Eurocentric, view of urban history that the age of large cities began with mercantilism and the rise of northwest Europe as the core of the capitalist world-economy from the sixteenth century on. This is not so. Earlier empires, focused on centres outside that small part of the earth's surface, developed very large cities, largely based on a combination of rank redistribution with some mercantilism. Few of these empires were very long-lived, and many of the cities declined substantially soon after attaining their maximum populations. While they flourished, however, the cities were, in the context of the available food-producing and transport technology, very large, drawing on extensive hinterlands and clearly based on well-organized systems of social control.

The evidence on city populations in the past is fragmentary and imprecise. What is available has been brought together by Chandler and Fox, in their monumental *3000 Years of Urban Growth*. They show that in the year 1000, no city in northwestern Europe had a population in excess of 40,000, whereas further south Cordova and Constantinople each had about 450,000, Seville 90,000 and Palermo 75,000. Three hundred years later, Constantinople had 150,000, as did Granada, and three Italian cities (Venice, Milan, and Genoa) all had at least 100,000. Palermo's population had almost halved. Further north, Paris had grown more than tenfold, to 228,000, but London still had only 40,000. By 1700, the northern cities were booming – London had about 550,000 and Paris 530,000 inhabitants, and Amsterdam, Vienna, Moscow and Lisbon were all growing rapidly; Constantinople remained the largest European city, however, with 700,000 residents.

The great cities of northwestern Europe are relatively recent phenomena, therefore, and in earlier centuries their size was far exceeded by southern European centres. Urban growth in the Americas followed that of northwestern Europe, with only two – Potosi and Mexico City – exceeding 100,000 residents during the period of Spanish rule. In North Africa, on the other hand, there was early urban development paralleling that on the northern shores of the Mediterranean: Cairo was estimated to have 200,000 residents by 1200, and Fez 250,000. Further east, Bagdad had 700,000 in 800 AD, Changan in China had 800,000, and Kyoto (Japan) and Hanchow (China) had 200,000 each. Some of these declined rapidly, notably Changan, but others grew to replace them: by 1000, the estimates suggest that twelve Asian cities had populations of 100,000 or more, and by 1300 there were sixteen. The first to reach estimated populations of 500,000 were Vijayanagar (southern India) and Peking. In 1750, the estimated population of the latter was 900,000, substantially larger than Constantinople, London or Paris: twenty-six Asian cities had 100,000 or more then, compared to seventeen in Europe.

Massive urbanization took place in pre-capitalist societies outside Europe, therefore. In 1200 AD only Constantinople (fourth), Palermo (tenth) and Paris (eighteenth) of European cities came in the top twenty-five cities of the world; a century later, Palermo had dropped out, but Venice, Milan and Genoa had joined the top twenty-five. By 1500, still only five European cities were in that group – with Naples replacing Genoa – and the number had only increased to six 200 years later (the cities were Constantinople, London, Paris, Naples, Lisbon and Amsterdam). Even in 1850, when the industrial revolution was well under way in the core of the capitalist world-economy, Peking, Canton, Hangchow, Bombay, Yedo (Japan) and Soochow (China) were in the top ten. By 1900, however, London, New York, Paris, Berlin, Chicago and Vienna occupied the top six places.

Urban-focused societies, with very large cities drawing on extensive hinterlands, are not peculiar features of European mercantile, industrial and late capitalism, therefore. Nor are they peculiar to the countries of northwestern Europe and their major colonies (notably in North America). Today, with the partial exception of China, the cities of the world are all nodes in an international economic system focused on a few countries. It must be realized, however, that – within their economic and technological contexts – other societies and modes of production have been able to support very large urban centres. Urban living then was undoubtedly experienced by a very small minority of the population only, but such was the organization of the societies that in absolute numbers the cities were very substantial indeed.

Summary

The focus of this chapter has been the urban system which existed prior to the introduction of industrial capitalism. In some places that system has been built on foundations laid down under rank-redistribution conditions, whose dominant characteristic has been a central place hierarchical structure. In most parts of the earth's surface, however, the major impress has been that of the mercantile socio-economic system, represented on the ground by primate urban centres dominating their hinterlands. The latter still dominate in many places; in almost all, the mercantile inheritance has provided the basis for an industrial capitalist urban system.

4. The Evolution of Urban Systems: II Industrialism

Mercantilism has been replaced as the mode of production in much of the world by industrial capitalism, a system characterized by the alienation of labour and by capitalist investment in work rather than in goods to be traded. Although, apart from those areas identified in Chapter 2 which are now organized under some form of state capitalism, industrial capitalism dominates the international economy, in many parts of the world mercantile operations are still the most numerous. Many of the major impacts of industrial capitalism, including those reflected in urban systems, are concentrated in a few countries only. As a consequence, mercantile urban features still prevail in many areas and the industrial urban system, in almost all cases superimposed on mercantile frameworks, is not a dominant feature on a world scale.

The factory system and location

Industrial capitalism involves a different set of employer-employee relationships from those existing in the productive sectors of mercantile economies. In the latter, both farms and workshops were characterized by employers and employees working alongside each other at similar tasks; in the former, ownership and control involve employers distancing themselves from their employees, so that as industrial capitalism expanded new forms of business organization evolved to replace those of mercantilism. These new forms affected manufacturing industry before agriculture in most cases; indeed, pre-industrial modes of organization still characterize large parts of the agricultural sector today in countries where industrial capitalism has long dominated manufacturing.

The first steps to industrial capitalism in many places involved the employment of labour in the home. Entrepreneurs would buy the necessary materials for a manufacturing process and distribute them to labourers who would work on them at home. The products would be later collected, and the price (usually based on the volume of production) paid; in many cases, the materials had to go through several separate processes, and were transported to several workers' dwellings, before they were ready for sale. This process was a slow and inefficient one: employers had little direct control over their labour force, in particular the amount of work that they did and the quality of their output. Greater control was gained by the institution of the factory, which involved bringing the workers to the work rather than vice versa. Factory production methods carry with them a variety of advantages over the pre-industrial and 'putting-out' systems.

Internal scale economies

Internal scale economies, or the economies of large-scale operation, involve employers making the fullest possible use of the resources available to them, that is, of men and machines. The aim, of course, is to make profits, and so the more efficient the productive process is, the lower is the cost of manufacturing each item relative to the price which can be received for it. Organization to achieve internal scale economies assists in the drive for greater profitability.

In a workshop with few employees, each individual usually performs all of the tasks involved in the production processes undertaken there. This involves the prior acquisition of a number of skills in an apprenticeship which is lengthy and expensive for the employer; the result is that the product is not cheap. Furthermore, the equipment used in each particular task within the productive process will probably not be required for much of the time (the same is true of the variety of skills learned by the workman). While lying idle, most equipment will deteriorate so that part of the investment in it earns no returns; the more equipment that is needed for a particular workshop, the greater the capital outlay and either the smaller the return on that investment or the longer the period involved in recouping the investment (perhaps paying off the debt incurred when the equipment was bought). To cut costs, and to increase potential profitability, equipment must be used more effectively and efficiently, and this in turn will probably mean more efficient use of labour. The factory system provided the means of achieving such a goal.

In the factory system each employee is given a single task only, and is responsible for only one piece of machinery (or group of related machines). In performing the one task only he can develop a greater skill at it, and his time does not involve 'wasted effort' moving from one task to another; at the same time the machinery is employed continuously, thereby reducing the fixed-cost component of the total production cost. (The fixed-cost component involves the payments for overheads such as machines and buildings, which have to be met whatever the volume of production. The other major component involves varying costs, which include the materials and labour employed. Clearly production is more cost-effective, given a standard price for materials and labour, the lower the fixed costs for each unit produced.) And if the employee is involved in a single task only, he will be less expensive to train since he needs many fewer skills than does the craftsman who performs all of the tasks in the productive process.

The movement from the craft workshop to the factory in which each employee undertakes a single task only was not achieved in one step; instead, a slow process, which is still developing in most industries, involving the increasing specialization of tasks, occurred. As this proceeded, and the division of labour became more fine-grained, so did the average size of factory. In a workshop the volume of production is small, but in a factory, if the machinery involved in each component part of the manufacturing process is operating continuously, the volume of production is much greater, so that many more people are employed in the various tasks. And because the factory-owner is in business to make profits, his continual aim will be to increase productivity. In part, as in all productive tasks, this may be achieved by demanding greater effort from the employees but, as already indicated in this book, this soon encounters the law of diminishing returns, in this case because of the volume of work which an employee can undertake (perhaps the number of machines he can mind) and the capacity of the machines themselves. The latter may be increased by the introduction of round-the-clock shift working, but this only delays the onset of diminishing returns. In the long term increased productivity can only be guaranteed by providing the tools which will make labour more productive. This means investment in newer and better machines, which produce more in a given period of time than do those that they replace. Such machines may involve the subdivision of the previous technology, involving even more machinery and more specialized tasks for the employees. And the greater the volume of production from each machine, the larger the size of the factory is likely to be.

Development of the factory system therefore involves greater specialization of tasks for the labour force, greater productivity and increased volume of output (which, of course, must be saleable). For many members of the labour force this specialization means that the skills required of them are few, especially when compared with their predecessors in craft workshops; the period of training is shorter and, because presumably more people are qualified for such tasks, wages can be held relatively low. Increasingly, too, machines take over tasks from labour, reducing that component of the production costs. Alongside this trend, however, will be an increasing demand for a relatively small, élite skilled group of employees involved in the design and maintenance of the machinery. The result is a polarization of tasks, skills and rewards within the factory, for this skilled élite will need to be highly trained and can command high prices for its labour. Other tasks will be created, too, such as those of overseers to control the productive process for the employers, because the factory is too large and complex for them to exercise the control personally, and those of office and warehouse staff to maintain a steady flow of inputs to the factory workers and of output from it. All of these tasks involve labour, who must be paid out of the surplus (the difference between 'making price' and 'selling price') and their wages add to the desire to increase productivity on the part of the factory-owner. (Some may be replaceable by machines, but the trend to automation in warehouse and office has not been as rapid, until recent years, as that in the factory itself.)

The factory system, therefore, is built on the drive for increased productivity of labour involved in the manufacturing processes. Such productivity carries with it the potential of increased profits for those who invested in the factory (its owner and those who advanced him capital) and it is largely based on the achievement of internal scale economies. It is not always the case that the larger the factory, the greater its efficiency, but this is likely to be so.

External economies of scale

Factories take in materials, process them, and put out a product. With some, all of the materials are in a 'raw' state – having been obtained directly from the resource base – and the products are ready for sale in the market. Such a factory is an independent production unit, except insofar as it is reliant on other factories for machinery. It is a relatively rare phenomenon, however, and increasingly so as industrial capitalism develops. Most factories require some of their inputs to be partly processed materials, which have already passed through other factories, and in addition they require some finished products from other plants (such as packaging); as a complement of this, many send some if not all of their output to other factories for further processing.

Most factories, then, are linked to others in complex webs of interlocking vertical chains along which products flow in their transition from raw materials into final commodity. The existence of such links indicates the interdependence between factories which characterizes the industrial capitalist system (especially in its earlier stages; see below p. 103). Such links may arise in a variety of ways. Factories set up to manufacture a certain commodity, for example, may find that they need inputs from five or six others, plus machinery from two more; some of their products may be purchased by other factories. Alternatively, a factory may develop as an offshoot of an existing one, specializing in the production of only one of the components fabricated there. The original factory may demand insufficient volumes of that component for it to enjoy the full internal economies of scale in its manufacture, but a separate factory, producing the component for a number of customers, may be able to. In this way the first factory can externalize some of its costs, if it is cheaper to purchase the component from the specialized producer than it is to make a small volume of it in its own plant. This is why the web of linkages interlocks. As the search for greater productivity continues, so the technological advancements which require large volumes of investment need large markets. Specialist producers are often best able to develop both the technology and the markets, serving a variety of customers. Just as the effort to achieve greater internal economies of scale involves increased specialization and division of labour, therefore, so also it generates new industries specializing in the manufacture of particular components. Each of these new industries is linked with a number of others, so that industrial capitalism is characterized by a complex of interdependent plants.

A further type of external economy relates not to links with other producers, who provide inputs and consume outputs, but rather to contacts with a range of service establishments; such economies are often termed 'urbanization' or 'agglomeration' economies. A factory, for example, will need its machinery maintained; some maintenance may be undertaken by its own employees, especially of machinery which it has lots of, but specialized pieces may have to be catered for by outside contractors. Similarly, it will have to transport its output, and perhaps its inputs too. It may do this in its own vehicles, but if only small volumes are moved, particularly over long distances, it may be more efficient to employ others who offer a transport service to a variety of producers. It may also need to use service establishments to organize its shipping, not only the actual movement but also the finance and insurance; it may need to employ accountants for particular financial tasks, or stockbrokers, lawyers, bankers and a range of other agents, simply because it does not provide sufficient work in any of these fields to justify full-time employment for a member of the firm. Similarly, it will need to purchase certain facilities and utilities, such as water, power, sewerage and garbage collection, roads, police and fire protection, and so on. All of these must be provided by either other firms or public agencies, which form part of the web of linkages that creates the industrial complex.

Factory location

The operation of industrial complexes involves the articulation of a great volume of transactions, some of goods, some of services, some of ideas. Individual production units in an industrial system are not linked with all others, however; most are linked to several, but have their main contacts with only a few. (For those with contacts with many, either for inputs or, more probably, for outputs, service establishments, such as wholesalers, may be set up to articulate the flows from many to many and avoid the movement of very large numbers of small consignments.) The articulation is a cost to the production unit, the firm operating the factory: the movement of materials and of goods in particular involves transport costs. The greater these costs are, the greater the proportion of the surplus (the difference between 'making price' and 'selling price', assuming that transport costs are not incorporated in the former) which is taken up in this way; as a consequence profitability is reduced by increased transport costs. Thus it is in the interests of the capitalist investors who are making decisions as to factory locations to minimize their transport bill (assuming that all other costs are held constant). Such minimization should be achieved by locating the factory as close as possible to all of those establishments with which it is linked. (This argument assumes that no firm can manipulate the selling price of its product. This is a reasonable axiom for analyses of the early stages of industrial capitalism, which are characterized by large numbers of buyers and sellers in each industry; it is not as reasonable for the later stages of monopoly capitalism, as will be made clear below. If the selling price can be manipulated by the producer, then it can be so fixed as to cover all transport costs.)

Because the links for most factories are several, it is impossible to locate them adjacent to all of their suppliers, servicers and customers. A choice must be made as to which it is most profitable to locate near, and this choice is likely to be influenced by the relative costs of transporting inputs to the factory, of moving outputs to the customers and of obtaining necessary services. In a simple example, a factory may obtain all of its inputs from **a** and sell all of its products to **b**; 10 tons of inputs from **a** are used to produce 5 tons of final product. If the former cost £10 per ton to transport per km, and the latter cost £7 per km, then the following calculations would apply: location at **a** would involve zero transport costs per ton for inputs whereas sales to the market at **b**, ten miles away, would cost £70 per ton – total transport costs per ton of output would be £70; if the factory were located at **b**, the transport costs per ton of input would be £100, those for movement of the output to the market would be zero, and, because each ton of output requires two tons of inputs, the total transport cost per ton of output would be £200; location halfway between **a** and **b** would involve transport costs for each ton of output of £100 for moving inputs (2 tons over 5 km at £10 per ton) and £35 for moving output, giving a total of £135. Extension of this analysis to all points would show that the closer the factory was located to **a**, the cheaper

TWO TONS OF INPUT (FROM a) ARE REQUIRED TO PRODUCE ONE TON OF OUTPUT (SOLD AT b)

TRANSPORT COSTS – INPUT: £10 PER TON PER KM; OUTPUT: £7 PER TON PER KM.

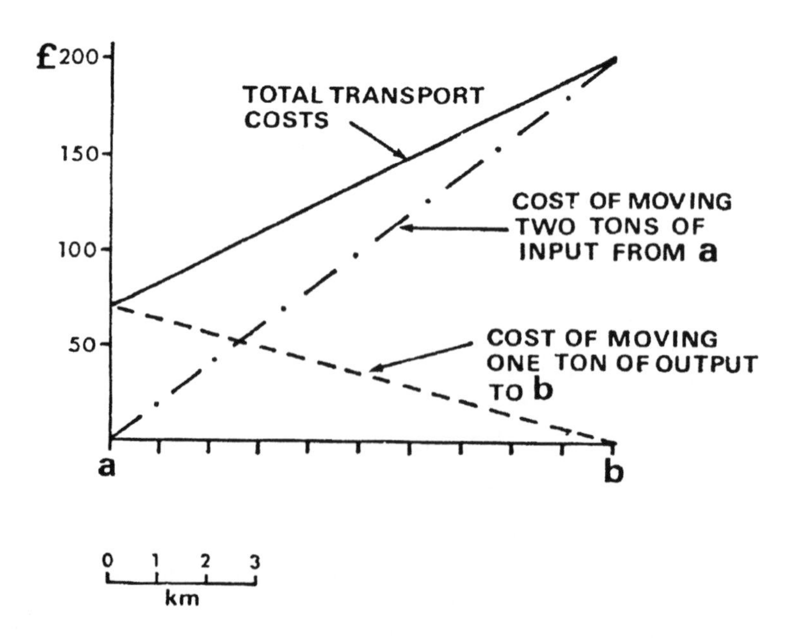

Figure 4.1 Diagrammatic illustration of the location problem with regard to transport costs.

the total transport bill would be; Figure 4.1 illustrates this diagrammatically. In this case, **a** is the better location because of the bulk of the input relative to that of the output, even though the latter is more expensive to transport; in general, the smaller the ratio of bulk of inputs to bulk of outputs, the more likely it is that the market point **b** will be the cheapest location for the factory. With many links, to suppliers and to markets, the more complicated the location problem becomes, and several locations (including some which are at neither a source of inputs nor a market for outputs) may offer the cheapest total transport bill. If the costs of other items in the total production bill, such as wages, land and capital, also vary from place to place, the search for the cheapest location becomes extremely tedious, though not necessarily difficult.

Few factories will be located in the cheapest places, of course, if only because of the fallibility of the human decision-maker and the likelihood that he has incomplete information on which to base his choice. (He cannot predict the future variation in prices, nor the behaviour of his competitors, with any certainty, and the consequences of his decision are likely to last a long time, because of the large investment which must be paid off once a location has been selected and a factory built. Because of this, he is likely to be conservative in his selection of a location and not gamble.) Thus maximum profits are not made by factory-owners; very bad decisions will probably lead to business failure, because of an inability to compete with those who can sell at much lower prices – see below. Each decision-maker will have a range of potential locations available to him at which he will be able to make a satisfactory profit (it is unlikely that he will know the boundaries of this range with any certainty); he may select the optimum location as much from chance as from design.

In general, for factories dependent on large volumes of bulky raw materials, or partly processed inputs, relative to the volume of their outputs, the feasible range of locations will be centred on the source of one or more of those inputs. For those further along the production chain, whose volume of inputs is relatively slight and which need proximity to customers (for a variety of reasons, including the perishability of the product), location near to major markets will be the more likely, although the feasible set of locations may be fairly large. A variety of other factors may attract such industries, often termed 'footloose' because of the small number of constraints on them imposed by transport costs, to particular places; as will be detailed in later sections, the increasing volume and range of linkages involved in advanced industrial capitalism and the declining dependence of industry as a whole on bulky supplies of raw materials make large cities, with their markets and urbanization economies, the most attractive locations.

The factory system and the urban system

A salient characteristic of the factory system is its association with scale and the employment of large numbers of individuals, compared to the system which preceded it. Thus the establishment of factories at a place generates the concentration of population there, both to provide labour for the initial factories and because the latter are likely to attract the establishment of others nearby, in order to benefit from external economies of scale. The result is the industrial town.

Where within existing urban systems are such industrial towns established? Since industrial capitalism grew out of mercantilism, and represented a reorientation of investment, it did not begin on an empty map. There was a population distribution already, served by a system of urban settlements, and industries were superimposed onto that system. Many were located in the existing towns, so that the urban industrial pattern was a modification of the central place hierarchies and gateway primate cities of mercantilism. Some, however, distorted that framework, with the creation of new industrial towns.

Industry and 'new towns'

Although it is convenient to write here of new towns established by the growth of industrial capitalism, in fact relatively few have been entirely new foundations. Most have been additions to pre-existing settlements; they are termed 'new towns' because the original places were relatively small (in terms of the mercantile system), and the onset of industrialism generated very rapid growth, moving the settlement rapidly up through the urban rank-size distribution.

Four types of industrial new town can be identified. The first comprises the *mining towns*. These are not truly factory towns, at least in their initial development, since their establishments are not involved in the processing of raw materials (although in many the required mineral is separated from its matrix of slack). Nevertheless, in the industrial capitalist era mining developed all of the characteristics of factory-based industries, such as the alienation of labour, the search for economies of scale and the introduction of new technologies to increase productivity. Some of the towns grew at settlements previously characterized by their mining activities, with the introduction of industrial technology generating population expansion. Others were created virtually out of nothing, as the result either of the discovery of a mineral which was in demand or of the increased demand for the product of a known resource. Their growth was often very rapid (as was the subsequent decline of many, when either the resource was worked out or the demand fell); capital had to be invested not only in the mine but also in housing for its employees (sometimes by the employer) and in a rudimentary infrastructure (shops and other businesses, notably drinking-houses) for urban living.

In the second type are the *heavy manufacturing towns*, containing the industries which process large volumes of bulky raw materials and which need, in most cases, to be close to the source of at least one of those materials. Coalfields were common locations for such towns, particularly coalfields which also contained resources of iron ore nearby and so were very attractive locations for that most basic of heavy industries, the manufacture of iron and steel. Many of them indeed grew out of mining towns, though the majority of coal-mining towns never gained a heavy industrial complex.

The third type contains the towns which attracted industries to a source of power for driving machinery. Many of these *power locations* were coal-mining towns but, except where it was needed in great volumes, coal could be transported to the user and so did not attract all types of industry to it. In early industrial developments the power potential of running and falling water, which could not be transported, was an extremely potent location factor, as along the fall-line on the eastern flank of the Appalachian Mountains in the United States. Modern technology allows this power to be transported as electricity via cables (as also can power generated from coal and other non-replenishable materials) but for industries which consume very large volumes of power (such as the smelting of alumina) location at or close to the source is still highly desirable, because of the loss involved in transport.

Finally in this categorization are the towns of the fourth type, those which grew up as *transport centres* and which, at least initially, had no major concentration of factories. In the areas where industrial capitalism originated, much of the long-distance movement of commodities was by water and, if necessary, a network of water routes (canals) was constructed for this. Port facilities were built, many of them at existing settlements, but some as the nodes for what became 'new towns'. For inland movements the canals were fairly soon superseded by the railways, and the port towns lost size and status to the railway junctions, some of which attracted industry either because of their centrality on the new transport network or to serve the railway system itself (as at Swindon, Derby and Crewe). For international trade overseas ports grew as the organizing focal points. Again, some developed industries associated with the port function (such as shipbuilding), whereas others were chosen as the sensible sites for the processing of imported materials (such as sugar, oil and iron ore). A wide variety of transport centres was established, therefore. Some never attained any major status or attracted large numbers of residents; some grew rapidly, but then declined as the major stimulus to growth was removed (a seaport whose harbour silted up, for example); a few rose to eminence within the national urban system (such as Liverpool); and some have developed particular industrial functions only (as at Teesside).

The establishment and expansion of these new industrial towns required the injection of large volumes of capital, in factories, in housing and in an infrastructure (shops, roads, utilities, etc.). Some of this capital may have been raised locally from those – landowners, merchants and craftsmen – who had profits from mercantile activity available for investment and were persuaded that industrialism offered a better rate of return on their money. In a sense, therefore, the more prosperous areas prior to the onset of industrialization were those best able to invest in, and capitalize on, this new form of productive activity, so that new towns were more likely to be founded in areas of mercantile prosperity. But large areas were scoured by the capitalists of the most prosperous areas, in the search for resources which were not available in their home regions. Capital moved in a colonial fashion, therefore, as it had in the mercantile period, and was invested wherever potential returns were perceived. In general, the search process proceeded outwards from the capitalist centre, so that resources near to hand were most likely to be exploited first and the new towns not established too far away. Thus areas distant from the main capitalist centres, and lacking local capital (because their mercantile development was as yet slight or because the profits from mercantilism there were flowing back, in colonial fashion, to the capitalist centres), received few new towns. By the time industrial capitalism spread to them, technological developments meant that few such towns were necessary, because there was less movement of bulky materials and power could be transported, so that industrialization was very much focused on the existing mercantile centres.

Industry and the mercantile urban system

The new towns created at the onset of industrial capitalism stand out in some urban systems because of their newness and the modification which they introduced to the existing urban rank-size distribution. But most of the industries established in this new era were in the existing towns and cities, where they served the established markets and generated further growth. In particular, they were the chosen locations for fabricating and assembly industries, which obtained partly processed materials from the heavy industries of the new towns and transformed them into products for their local markets. (Not all industries in the new towns were of the 'heavy' variety; the early mechanized textile industries needed power supplies which in most cases could only be obtained in locations lacking major urban settlements – as on the flanks of the Pennine Mountains in England.)

Three major factors account for this concentration of many factories of the industrial era in the pre-existing mercantile centres. Firstly, the established towns had capitalist traditions, of entrepreneurship, investment and lending, and these could be moved into the industrial sphere. (In part, this might produce a movement of capital away from trade, but that would be relative and not absolute in its volume. Industrial capitalism required trade, so successful industrial investments would not only create the bases for trade but would also generate the capital which could be invested in it. The two were interdependent, and were both stimulated by the development of the modern banking system, which generated an increase in the available money-supply.) In addition, the craft industries of the mercantile towns formed the basis for many industrial establishments, as their owners changed their mode of production and invested in factories and machinery, using some of their own capital and accumulated profits. Finally, the trading tradition meant that these towns had access to local markets which had been fostered by their traders, and which provided ready outlets for local production. As a result of these three factors, therefore, those market-oriented industries which were not tied to particular locations by their demands for either raw materials or power created industrial complexes in the settlements inherited from the mercantile era.

The established pre-industrial towns were the articulation points for the mercantile trading system, which gave them a further advantage as industrialization proceeded. The trading functions themselves, most of which were expanded as a result of industrial growth, generated demands for other industries necessary to the increased volume of trade. Further, the capital available there because of the success of trading ventures could be used to finance not only industry but also transportation improvements; the latter stimulated growth in both the industries that provided the materials for the transport systems and those which produced goods to be consumed in the hinterland. Such investment furthered the superiority of the pre-industrial towns relative to that of competitors which were not so fortunately placed, and some were able to support not only industries that served the commercial sector, either directly or indirectly, but also those which were entrepôt-based, processing materials obtained from one source and shipping the products to a market much wider than the local.

The central place settlement pattern inherited from the mercantile era thus had a hierarchy of industrial functions grafted onto its hierarchical trading organization. Most of the new industries served only the town and its immediate hinterland; a few produced for larger areas, with the distribution of their products being articulated through the central place system. Such market-oriented industries had critical thresholds, just like those for trading functions: minimum populations necessary to provide the demand which would support a profitable factory. (Such thresholds could only be estimated as averages, of course, since there would be place-to-place variations in, for example, various costs such as labour and raw materials and the demand for the product.) The smallest towns would be able to support only a few factories, providing commodities which were in great demand and which, in several cases, were not transferable over more than a short distance; breweries, bakeries and brick-making plants would be typical of such smaller settlements. Few of them would be organized in industrial complexes; most would obtain their machinery from other places (probably the new towns). The bigger towns, on the other hand, would meet the thresholds for a larger range of industries – serving other, smaller settlements within their marketing area as well as the resident population – and would be more likely to develop small industrial complexes.

The general relationship between urban size (as measured by population) prior to industrialization and the number of factories established there is probably an exponential one: if town X were twice the size of town Y, it may have been able to support the establishment of three times as many factories, because Y's population meant that it crossed only a few of the significant thresholds for industrial location. As a result, X's population would increase relative to that of Y, via the larger working population that it attracted from the rural areas to its factories, thus widening the gap between the various levels of the urban hierarchy. (If there were no gaps, of course, the result may have been to create some.)

If, as suggested in the previous paragraph, the result of the introduction of the factory system to a mercantile urban system was to accentuate the size and economic importance of the larger centres, this should have been most apparent in the areas dominated by gateway primate cities. These places would attract most of the industrial investment because of their centrality to the local market, their relatively large populations and, in most cases, either their status as ports or their positions as major transport nodes which would assist in the import of materials and technology. Thus in Australia, for example, the industrialization which has accompanied the achievement of post-colonial status in the twentieth century has been focused almost entirely on the gateway primate cities. And so in Victoria the only inland centres of any size, the former gold-mining towns of Ballarat and Bendigo, were in decline by 1960 and in New South Wales there has been no substantial development of industry outside the Sydney metropolitan area, with the exception of two new towns based on coal and on iron and steel at Newcastle and Woolongong.

Urban growth

In general, therefore, the onset of industrialization resulted in an expansion of the existing towns; and the larger the town, the greater the size of its industrial sector and the larger the immigrant population which it was likely to attract. Some new towns were established, to cater for industries with particular requirements that could not be met in existing settlements. The earlier the onset of industrialization in a system, the greater the number of new towns founded, because the technology used in the early decades of the industrial revolution required large volumes of raw materials, was unable to transport bulky materials over long distances and could not move power. As investment was used to make the industrial processes more productive through technological changes, so locational constraints were reduced, except for a few so-called heavy industries. Textile manufacturing, for example, was initially dependent in the early decades of industrial capitalism on substantial volumes of water power such as could be obtained on the flanks of the Pennines in northern England and along the Appalachian fall-line in the eastern United States. Increasingly the industry has become more footloose, however, allowing its development in other areas of those countries (as in the American South) and, in other countries, for the establishment of factories in the existing towns rather than necessitating the foundation of new towns.

In all towns, industrialization developed through a self-propelling growth process, which reflected the complex of inter-industrial linkages;

the volume of this growth has usually varied according to urban size. Take a factory which is newly established, and provides 1,000 new jobs. The workers taken on to fill these vacancies have families, and so the total population affected may be about 4,000. They may need homes close to the factory, which creates jobs in the construction industry (as did the building of the factory itself), and also in the industries which serve it, such as brick-making and timber preparation. This may involve a further 100 jobs, supporting 400 persons, who themselves may want homes. So the creation of 1,000 new jobs may create the need for 110 more and support a population of about 4,440. These people will all be consumers, and their demands will create employment both in the shops from which they buy and in the industries, of which there may be many, which serve those shops. Thus growth in one industry has stimulated growth in many others, which may include itself. (Expansion of a steel mill, for example, may demand more iron ore, and expansion of the mines to meet this demand needs steel with which to construct the new tunnels and lifting-gear, plus the railways to move the ore.)

And so the process continues. As long as one industry is growing, others will be too, since no industry exists in isolation from others. Indeed, although most industries are not directly linked to all others, almost all will have an impact on every other one when they expand, because of the growth in consumer demand which this generates. (The reverse happens with decline, of course; a fall in production in one factory will affect the linked factories, will lead to redundancies among the affected workforces, and will lead to reductions in consumer demand, thereby influencing most other industries.) The growth in one factory or industry need not create additional jobs in all cases. It will create extra demand, but this can be met either by employing more workers or by increasing the productivity of existing workers. Whichever occurs, the amount of income circulating is increased, which produces greater demand for goods and services, some of which will probably be reflected in job opportunities (especially in the service sector).

Creation of new jobs or wealth generates even more of the same, therefore, through the demands of the industry concerned for inputs and through the needs of those employed for goods and services. Together these comprise a process which is usually known as the *multiplier*, whose nature is represented diagrammatically in Figure 4.2. The growth in this example begins in industry X, and the jobs created stimulate a growth in population. Industry X gets its inputs from other industries with which it is linked, and their expansion, consequent on the growth of X, both stimulates further population growth and makes demands on those

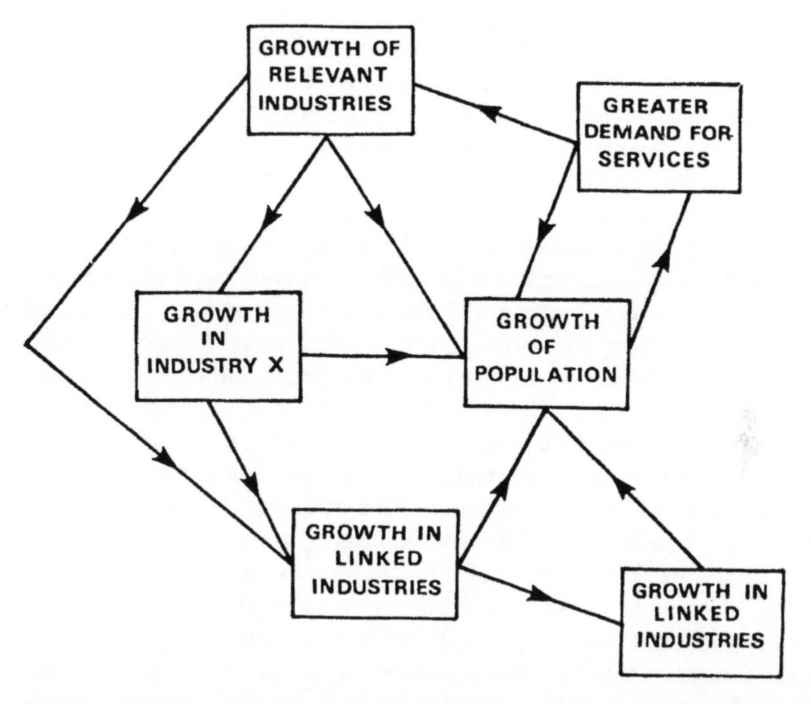

Figure 4.2 Diagrammatic representation of the multiplier process, illustrating the various effects of growth in industry X.

other industries with which they are linked. The population growth associated with this linked industrial expansion leads to a demand for goods and services, which must be met in the trading sector and which in turn generates further industrial growth (which, as suggested in the diagram, may make demands on both the industry in which the growth began and those with which it is linked). Since each new job in the original growth industry usually generates less than one job in those with which it is linked, the circular process of growth eventually dies out, the extra demand created being insufficient to justify further employment expansion.

All of the growth identified in Figure 4.2 is based on the original expansion of industry X. Thus growth in X in that case is basic to all other growth, which leads to the designation of X as a *basic industry*: all the others are dependent on X, and so they are termed *non-basic industries*. To some extent, all manufacturing industries are basic, although many (such as the manufacture of machines) are very much dependent on others. Similarly, the amount of extra growth which may be generated by expansion of a basic industry (termed its basic:non-basic ratio) varies considerably from industry to industry. This has been shown by detailed studies of the links between all pairs of industries in an economic system (the representation of which is known as an input–output matrix). Certain industries have much more impact on an economy (both positively in the creation of jobs and wealth when they expand and negatively in the induction of unemployment when they decline) than do others, and it is the identification and fostering of such industries which is central to the economic planning of the state in late capitalist societies.

So far, the location of the multiplier effect has not been considered. In general, the larger the town in which the original expansion takes place (industry X in Figure 4.2), the more likely it is that the extra expansion which is generated is internalized there, and so is non-basic in a locational sense. If, for example, a new factory is opened in a small town, the linked factories which provide its inputs and receive its outputs are likely to be elsewhere, so that the links are external to the town and the extra employment is generated elsewhere. Similarly, the goods and services bought by the workforce and their dependants in the shops of the town probably have to be brought in from factories in other towns, so that again the non-basic links are largely external to the place in which the expansion occurs. The growth generates more growth, but the local economy does not benefit very much from it.

In larger towns, on the other hand, the factories serving the basic industry that is expanding are more likely to be local, so that the employment generated is internal to the settlement. The same is true of the industries providing the goods demanded in the shops by the expanded workforce. Thus the larger the town, the greater the volume of the non-basic growth which is retained there (which is reflected in the size of the basic:non-basic ratio for the settlement itself rather than for the industrial system as a whole; some growth will be external, however, perhaps even to the industrial system). Furthermore, if the requisite linked industries are not present in the town, the new demands there might be sufficient to justify the establishment of a factory to meet them, thus increasing the basic:non-basic ratio there. In sum, the larger the town, the larger its basic:non-basic ratio and the greater its ability to internalize any local growth impulses.

Again, therefore, it is the larger settlements which are likely to benefit most from the growth associated with industrialization, since they are able to meet larger proportions of the non-basic demands created by any basic expansion. The larger the town, the more the growth impulses are internalized and the multiplier circulates locally; the smaller the place, the greater the proportion of the benefits from expansion there which are enjoyed elsewhere. Thus in urban systems which comprised central place hierarchies prior to industrialization, the hierarchical format is likely to be maintained and the gap between the settlements at two adjacent levels is likely to widen. In gateway primate systems, the hegemony of the major cities will be extended, probably considerably.

New towns established for particular industries will probably be at a disadvantage relative to older places in the internalization of multiplier links, because they lack the non-basic establishments. Like the small towns (and many new towns remain small, of course), most of their growth impulses will be transferred elsewhere. Occasionally a new town may grow to large size, because of the demand for the product of its basic industry; this may lead to the development of a non-basic sector, particularly if the profits of the basic industry remain to be invested locally in the consequences of growth. New towns like Birmingham (England) and Pittsburgh (U S A) are good examples of this. In other cases, one town in a group of new centres may develop administrative and other functions to serve the region and will grow accordingly, as in the centres of the textile regions of northern England (Leeds and Manchester): these have grown to sufficient size, based in part on the prosperity of the whole region, to internalize much of the multiplier.

Innovation and growth

One further aspect of industrial expansion which benefits the larger towns and cities disproportionately is the innovation cycle. Industrial capitalists are continually seeking new ways to increase the return on their investment, usually either by the invention of new machines and processes which will increase labour productivity in their factories or by the development of new products for which a market exists (or can be created). Within any one firm, this can be achieved by the creation of a research and development department, which absorbs some of the current surplus as an investment in the search for a greater surplus tomorrow; in general, the larger firms, enjoying considerable internal economies of scale, should have always been the best able to provide this investment and so secure future economic viability. In many instances, however, the inventive, speculative activities have been undertaken by small firms, owned and operated by scientists and engineers (some with no professional qualifications) who have invested their own capital, plus any which they have been able to borrow, in experimental processes and products. Most of them have been located in the larger towns, where there is both a critical mass of such people interacting to produce ideas and a potential market. Their success rates can be illustrated by analyses of where patents are taken out; in most urban systems a disproportionately large number, relative to their proportion of the population, has been filed in the largest centres. (Many of the firms fail, of course, either to invent anything significant or to survive as viable enterprises.)

Somewhat similar to this innovation process is the seed-bed process. Within a capitalist system it is the ambition of many employees to set up businesses of their own; the ethos of the self-employed, successful man is a very strong one. Again, the bigger the town and the larger the number of factories there, the greater the probability that one or more employees of an existing firm will set up on their own, using a combination of personal savings and loans. They take with them expertise gained in their previous employment; their aim may be to compete in the same field, although some will specialize in esoteric products for which there is a small market that lacks the large potential to attract investment from a big firm. Many of these small firms fail, often very soon after their foundation: their products may not be saleable; they have insufficient capital; they cannot produce cheaply enough; or their owners are poor businessmen. A few succeed, however, and expand their operations, thereby generating further growth via the multiplier processes. Success is most likely in a 'growth industry', one with an expanding market for its products. An example of this has occurred in the Santa Clara Valley of California during the 1970s with the manufacture of microprocessors using silicon chips. Initially there was one firm there (Fairchild), but several of its employees (known as the Fairchildren) have left to establish their own factories in this boom industry, forming a cluster of similar establishments (the area is now known to some as Silicon Valley).

There are several components to the economic multiplier in addition to those identified in Figure 4.2, therefore. These can be termed the service industry, the linked industry, the innovation and the seed-bed loops, each of which is linked to the population growth component, and thence to each other; their general form is shown by Figure 4.3. The links shown there are only representative of all the possible ones; both innovation and seed-bed growth, for example, are shown with links only to those industries which are linked to the basic sector, whereas both might feasibly be associated with either or both of the basic and the service industry sector. As before, the larger the settlement, the greater the probability that all four of these cycles are internalized, so that their growth impulses generate further expansion there rather than elsewhere. Some major inventions are achieved in small towns, of course, thereby generating

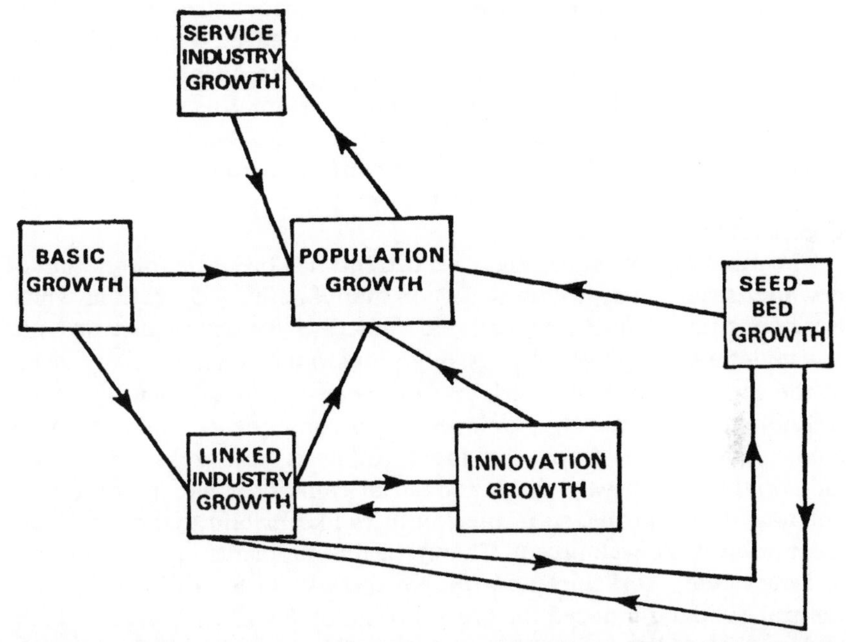

Figure 4.3 Diagrammatic representation of the four components of the multiplier process.

expansion which allows them to move up the urban system substantially, as has occurred in several places involved with the car industry (Detroit and Oxford, for example). Small places have relatively few industries, however, which not only reduces the potential for innovation, seed-bed growth and internalized multipliers but also means that successful entrepreneurs may find it necessary to transfer their businesses to larger centres. The nature of the local industrial structure may be relevant to the cycles of Figure 4.3 as well: Birmingham, with its mass of small firms in a range of industries, has a better innovation and seed-bed record than does Manchester, for example. And finally, gateway primate cities, from which many profits are repatriated to overseas owners of trading and industrial companies, may have a relatively low potential for such growth impulses, even if they are large.

Urban systems under industrial capitalism

The conclusion to be drawn from the previous sections, therefore, is that the main impact of industrial capitalism on urban systems has been to exaggerate the differences between places of various sizes that had been established in the preceding mercantile era. In effect, the gaps between the different levels of the urban hierarchies were opened up, with the larger places benefiting disproportionately from the various growth impulses that characterize industrialization. In particular, the primacy of gateway cities in many parts of the world has been accentuated. (In technical terms, this means that the rank-size relationship shown in Figure 1.1 has been steepened.)

The main exception to this general trend has been the foundation of new industrial towns, to house certain manufacturing operations which were constrained in their locational choices by access requirements to raw materials and power sources. Such constraints were especially severe in the early stages of industrialization when production and transport technologies were poorly developed. As a consequence, the new towns made their greatest impact in those countries and areas which were industrialized relatively early. Industrial capitalism had its origins in north-western Europe, so it was countries like Britain that experienced most new town development. There, the developments in the Midlands, in Lancashire, Yorkshire and the North-East, in South Wales and in Central Scotland induced the growth of many towns from a small base, and generated alterations to the country's urban system which remain to the present. By the time of the first census, in 1801, many of these new towns had been established and were growing fast. Nevertheless, during the succeeding century many of the long-established central places which did not attract major industrial functions (such as Bath, Exeter, York, Norwich and Chester) moved down the rank-size ordering while new towns (such as Oldham and Leicester) moved up (Figure 4.4).

Outside north-western Europe industrialization had less impact on the composition of the urban system, although it induced much growth. New towns were founded in the eastern United States, in Appalachia, on the shores of the Great Lakes and in parts of New England. But as settlement moved west, so it was the gateway primates which dominated the urban scene, and places such as Chicago and St Louis, Los Angeles and San Francisco attracted most of the new industries. Some new towns were established, mostly for mining operations west of the Rocky Mountains divide, but none grew to the size which generated multipliers and further growth; most are ghost towns now. And the same is true of most of the rest of the New World where, with the exception of areas whose basic industries are mining and iron and steel (such as the Damodar Valley west of Calcutta), the dominance of the gateway primates has been accentuated rather than challenged by the superimposition of an industrial system onto mercantile capitalism.

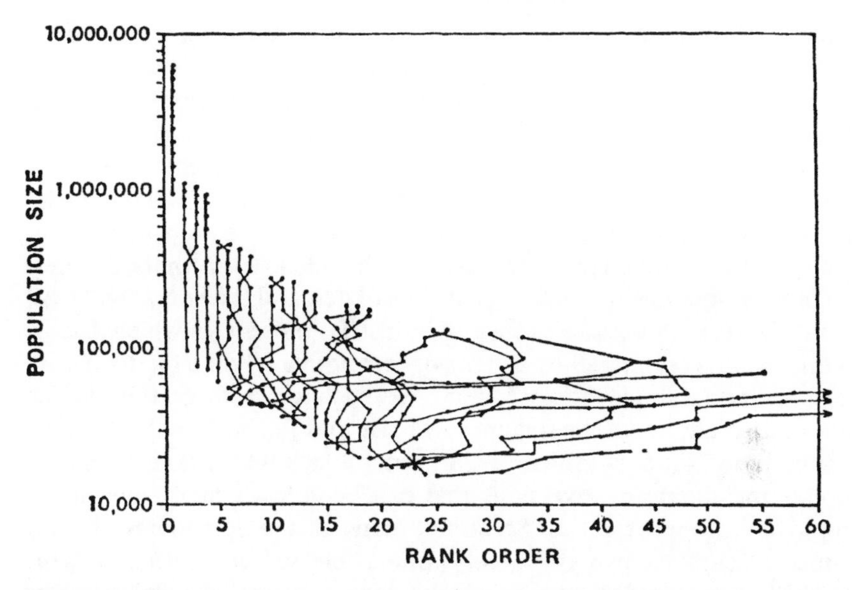

Figure 4.4 Changes in the rank-size ordering of towns in England and Wales: 1801–1911; only the twenty-five largest towns in 1801 are indicated. (Source, Robson, 1973, p. 391.)

The onset of late capitalism

Industrial capitalism is far from static. The drive for profits means that capitalist investors are constantly seeking new ways to increase their returns, to ensure markets for their products, to obtain access to raw materials, and so on. This encourages their investment in research and development, for example, the results of which are seen in the range and type of products offered for sale.

Several features of the dynamic of industrial capitalism are of relevance to the study of urban systems. The first is the change in the product mix, and consequently in the occupational and industrial structure, which follows from the introduction of new products. Thus coal has been replaced, in part in some places but almost completely in others, by alternative fuel sources, notably oil, gas and nuclear power. Steel, too, is in less demand, as other metals and synthetics are introduced to replace it, and the traditional raw materials for textiles, wool and cotton, are being substituted by chemically derived synthetics. Thus coal mines and steel mills are either reducing their employment to counter the falling demand for their products (some reduction would occur in any case because of productivity increases, of course) or are being closed completely. This sets off the multiplier process in reverse (Figure 4.2), as unemployment in a basic industry generates the same consequence in those with which it is linked, both directly and indirectly. Just as growth creates growth, so unemployment can generate more unemployment.

Two types of town are most likely to be affected severely by the operation of the multiplier process in reverse. The first are the small towns, dependent on only a few factories for their industrial livelihood. These are more vulnerable to changing patterns of demand than are their larger counterparts with a greater range of industries present, in which decline in one may be countered by expansion in another. (Some big towns are dependent on a single industry, of course, such as Seattle on the production of aircraft, and their fortunes will rise and fall with that industry.) Furthermore, the small towns will in general lack both the accumulated capital and the innovative skills that can be used to create alternative employment opportunities. Secondly, many of the new towns of early industrial capitalism may be very vulnerable to such shifts. A large number were founded on a particular local advantage, such as a raw material or power resource; if this advantage no longer obtains, then the town's basic industry will be in decline, and if no others have been attracted there, decay may be as rapid as the earlier expansion.

A second feature of late capitalism is the growing sophistication of production technology. As the basic needs of families for food, shelter and clothing are met (at least for a majority of the population), so the demand for these tails off and capitalists must seek to produce other types of goods, which consumers will not only buy but will also replace frequently. Materialism is basic to late capitalism. Without it, factory products (which are increasing through productivity improvements) will not be sold, insufficient surplus will be obtained, investment will fall, and unemployment will result. To feed materialism, more and more complicated products are devised and marketed: televisions with remote-control units; digital watches with calendars; calculators with built-in tape recorders; sophisticated games; new fashions in clothing; and so on. A characteristic of many of these products is the relatively low cost of the raw materials; the main element in their price is in the labour and machine power involved in the manipulation of the materials (much of the labour involved in the actual production process is unskilled and relatively cheap; the expensive labour is involved in the research and development and in the design of the necessary machinery). The industries involved are thus known as high value-added industries, value-added being the term that describes the difference between the cost of the raw materials and the price of the final product. Few are tied to special locations by their demands on certain resources: most are leading members of the innovation and seed-bed cycles identified in Figure 4.3. Because of their manifold links with other industries, their dependence on 'risk capital' in their early development, and the uncertainty of their markets which demands access to the means of achieving sales, most of these high-value-added industries are found in the larger towns.

The industrial developments of late capitalism tend to favour the large cities, therefore, and to accentuate the importance, both economically and in population numbers, of the major nodes in the central place hierarchies and of the major gateway primate cities. The small towns and the new towns of industrial capitalism fare less well, although some towns, notably those relatively accessible to the larger centres, may be attractive locations for factories involved in the unskilled operations of the high-value-added industries. In general, therefore, late capitalism is characterized by a retreat from the new towns and a concentration in the larger places. The former, it would seem, are a transitory phenomenon, typical of a certain stage of capitalist development only, although their fabric remains long after their utility to industrial capitalism has largely waned and their population as a result suffer large-scale unemployment (as in Glasgow and Liverpool). The trend is a strong one, however, and is accentuated by two other features of late capitalism.

Concentration and centralization

Although the two trends just discussed are important aspects of late capitalism, much more crucial for the urban system are the twin characteristics of concentration and centralization, whose major features were identified in Chapter 2. Concentration refers to evolution in any one sector of productive activity, and is a consequence of competition; as this proceeds, the weaker firms are eliminated, until eventually only one or a few remain, so that whereas the early stages of industrial capitalism are characterized by large numbers of relatively small, independent firms in each productive sector, in late capitalism most sectors are dominated by a few firms only. Centralization refers to the growing links between sectors of the economy, as the big firms from different sectors are merged to form large conglomerates involved in a wide range of unrelated activities (both productive and non-productive). Thus control over the economy becomes centralized into relatively few hands and, as a consequence, relatively few places.

Concentration and urban systems

Concentration was the first of these trends to make a major impact on the urban system; its operation involved important spatial influences and consequences. To increase the volume of surplus being produced, and thereby satisfy investors (who otherwise would starve them of capital), it was necessary for firms to increase their sales. Most firms, excluding those in the new industrial towns, would probably be serving a local market only from their factory, comprising the town population plus that of the immediate hinterland. If this market were not already saturated, it could be developed, but if it were, and the firm was unable to diversify its product range, the only ways in which it could expand involved either colonizing new markets not currently purchasing the product or competing with similar firms either in adjacent market areas or in their own.

Although colonization of new markets did occur, this was usually based on successful application of the second strategy. In order to compete in other markets in the urban system, access was crucial, and widespread concentration was delayed until transport technology had developed sufficiently to allow bulk movement of commodities. Once this occurred, the competition was based on the price and quality of the goods on offer. Regarding price, it was generally the larger of the two competitors which succeeded, because it was achieving the greatest internal economies of scale and so had the lowest unit production costs. Even given the transport cost disadvantage relative to the local producer, the larger firm could often sell to its market at a lower price, and thereby win over sufficient custom from the local firm for the latter's operations to become unviable. If necessary, the larger firm might sell in the other market at a loss on its unit production costs plus transport charges while seeking to defeat the local firm, covering its losses by profits elsewhere and accumulated capital, in the expectation of greater returns – from increased prices – once the battle had been won. The smaller firm may have responded by cutting its price and, perhaps, appealing for local market loyalty. But in most cases it eventually had to concede; the price competition reduced profits to intolerable levels, capital sources dried up, and the only solutions were either to go into liquidation or to be taken over by the larger competitor.

The large firm succeeding in its invasion of another firm's market area in this way is likely to have been based in a relatively large centre, certainly compared with that of the firm which it fights. Large towns offer larger local markets, with the associated internal economies of scale, plus greater external economies than are available in smaller places, and together these allow production costs which are often significantly lower than those in smaller towns: once transport costs began to fall substantially, so that they were less than the production cost differential between the large-town and the small-town firm, the former could begin its invasion of the latter's market. Thus concentration generally resulted in a spatial concentration of productive activity in the larger centres of an urban system, as the process of horizontal integration meant the closure of small-town factories. Where the small-town firm merely closed down, closure was inevitable; if it was taken over by its larger competitor, however, although closure was likely as part of the rationalization programme of the expanded firm, the local factory might be retained as a distribution depot, with substantially reduced employment. (Initially, larger towns may have contained several firms in the same sector, and the first stage in the concentration process involved one of these achieving local hegemony.)

Not every competitive thrust resulted in victory for the large-town firm and closure of small-town factories. Some firms in the smaller centres were able to maintain a hold on their local markets, and even to penetrate those of their large-town counterparts, through combinations of good management, better application of technology and good fortune. And the more successful they became, the more they were able to guarantee continued growth, both for themselves and for the towns in which their operations were centred. Increased accessibility, especially with the development of road transport, and the reduced dependence of factories on material inputs with the growth of technology, have improved the probabilities of small-town firm success, particularly if capital and enterprise are available. The towns involved are not usually the smallest in the urban system, of course; rather it is the medium-sized places which may contain the sorts of firms which are able to survive in the competitive world of late capitalism.

The new towns of the industrial capitalist era were likely to remain small, as pointed out earlier, and not to develop much of an internalized multiplier. Their fortunes were closely linked to their basic industries, and as these declined in importance in the overall production structure, so did their host communities. The processes of concentration were likely to occur in those industries as well, however, as a few firms succeeded in commandeering the market to themselves, with the consequence that spatial concentration occurred too; the results of this are seen in, for example, the closure of small coal mines and iron and steel plants and the concentration of both types of activity in a few centres only.

Concentration has operated in the service industries as well as in manufacturing. In retailing, for example, successful firms have expanded and invaded the market areas of others, usually by establishing branch stores in the relevant towns, by taking over formerly independent businesses in those places, or by encouraging shoppers to travel to larger places. The first two processes are made possible by the internal economies of scale achieved through size of operation (there are major benefits to be gained, for example, in the bulk buying of goods from their producers). These allow reduced prices to be charged relative to those of the local stores (most of which are independent businesses), eventually forcing the latter out of business. Again, it is usually firms in the larger towns which can operate this process most successfully. Eventually, they may be able to gain monopolies in the smaller towns, if they own shops there, which allow them to increase prices and profits. Alternatively, they may close down the shops in the smaller towns, obliging the local populations to travel to stores in the larger centres; this trend has been encouraged by increased customer mobility and preference for shopping in larger centres, with their greater range of goods and services.

Concentration in retailing and other service industries, and its results in the concentration of service provision in the larger settlements, parallels the concentration process in manufacturing industries. One of its results is that many small settlements are either denuded of their retailing sector or retain only a few small establishments which can compete with those in the larger towns in terms of convenience and access only; a consequence of this is that the relatively immobile, who cannot travel to the larger centres, probably have to pay higher prices for their goods.

A further concentration process parallels those in retailing and manufacturing, and in part encourages them. This takes place in the farming industry. Through the industrial capitalist era much of this industry retained many of the characteristics developed under mercantilism, and its organization remained typical of mercantilist craft industries. But to improve productivity farming had to become more capitalist in its organization. Increasingly, labour was replaced by machinery and concentration proceeded, as successful farmers, usually those with large landholdings, expanded their operations by taking over neighbouring farms. Both of these processes generated negative multipliers for the small towns which provided services to the rural areas, since both resulted in a reduction of the agricultural population, and thus in a decline in demand for the goods provided in the local stores. Local industries servicing the farming community also suffered, and with the increased mobility that encouraged the successful farmers to travel further and to trade with the large-town firms, the viability of many small settlements was destroyed.

These concentration trends have built up slowly, accelerating in recent decades with the rapid advances in transport technology. There has been no wholesale concentration into the larger towns, therefore, but rather a slow process of concentration of economic activities in successively bigger settlements. The hegemony of the largest has been ensured, especially that of the gateway primates, very largely at the expense of the lowest rungs of the central place hierarchies (with clear social consequences for the communities 'left behind').

There have been exceptions to this spatial concentration of economic and social activity. Those smaller settlements which have survived, even grown, have done so because inhabitants have been able to capitalize on two main assets. The first is a resource base attractive to a more mobile population. Scenery and community size are the main ingredients; they result in resort centres, in the establishment of colonies of second homes for city-dwellers (often not approved by the locals), and in the establishment of dormitory commuter settlements some distance from the major workplaces. The second resource is accessibility. With the improved transport technology industrial complexes can operate with their constituent parts separated by substantial distances, especially if the materials being moved are small in volume relative to their value. Thus small towns near to major industrial centres have been colonized by firms seeking pleasant locations for their productive activities (as well, in many cases, as cheap, non-unionized labour). Whereas on the larger scale the general trend is concentration, therefore, locally deconcentration (or sprawl) is common. Growth in late capitalist societies occurs in urban regions surrounding major urban centres rather than in the large cities alone.

Centralization and urban systems

Concentration can proceed only so far. Once a market has been captured from a competitor and saturated, the growth potential for the firm concerned is limited, if not ended. Fresh outlets may be sought in other countries and urban systems, in which case the process of concentration may lead to the repatriation of profits to the home country of the expanding firm. A result of multinational concentration, therefore, is the removal of some growth impulses from the countries in which they are generated; the countries that suffer in this way are usually relatively small (in population though not necessarily in area), so that multinational industrial organization has all the characteristics of the earlier phases of mercantile colonialism.

Limits to multinational concentration are set by governmentally imposed constraints on the flow of capital and profits, constraints which are aimed at retaining capital in the country of its generation, to provide the investment for the next round of development. Thus for many firms the desire to expand and invest can only be accommodated by centralization. This may involve vertical integration within a particular industrial complex, with a firm operating at one stage in the production process linking both its suppliers and its customers (other firms) into a single conglomerate. This may be accompanied by factory closures as part of a rationalization policy, but most of the plants are likely to remain unless major external and internal economies of scale can be achieved by a restructuring of the productive processes. One usual consequence, however, will be the channelling of the surplus produced by the various component parts of the conglomerate to the headquarters' location, which will probably be one of the biggest urban centres within the system, and this will generate further employment in that location in jobs involved in coordination of the larger organization. (One of the arguments for centralization is that it produces more efficient organizations, but there is little evidence that large firms employ fewer 'organizers' per unit of output than do their smaller counterparts.) To the extent that the firm is a public one, and is financed by public subscription to stock and share issues, its profits may go to any part of the system; the places with the greatest concentration of highly paid workers, however, are in general those with the largest volumes of investment in such companies, so that the income which creates both further investment and growth stimuli is disproportionately concentrated in those urban centres which contain the bulk of the headquarter offices of large, often multinational, conglomerates.

Vertical integration may offer some advantages, in addition to the purely financial, but the potential returns may be such as to convince a firm's investment officers to direct their capital elsewhere in the industrial system (or even out of the productive system, into, for example, land speculation). The results of such centralization activity may include factory closures, but only where the intention of the takeover includes asset-stripping. But by redirecting the profits from a factory away from its previous local owners, centralization will stunt the growth impulses in the towns where its new factories are. Thus profits from basic industries in one place may be used as expenditure in the service establishments of others, transferring part of the urban multiplier to the cities containing the headquarters of the conglomerates and housing their main investors.

Together, concentration and centralization lead to a growth in the economic dominance of an urban system by a few very large firms whose headquarters are concentrated spatially in a few large places. Within the firms there is some tendency for spatial separation of the various component parts: production and assembly; research and development; routine administration; and decision-making. The first of these, although probably concentrated in the larger settlements, may be distributed widely through the urban system, reflecting both the locations of the surviving factories which were taken over and the recent trend towards deconcentration in the location of factories employing largely unskilled and semi-skilled labour. Research and development functions are locationally footloose; they are usually concentrated in pleasant environments and close to linked institutions, such as certain universities. Routine administration, too, can be located in many settlements, large and small, as long as there are good communications with the other components (both telecommunications and transport routes); much of its operation requires relatively unskilled non-manual labour. Finally, the decision-making function in a multinational conglomerate may be located in another country, with only a branch office within the particular urban system. This, like the head office of a national firm, is likely to be in the political capital, with access to government offices, to the vast service infrastructure of such places, and to its competitors (many firms are interlocked at the directorate level, and so are not truly competitors): in most countries the capital city is one of the largest cities, if not the largest.

Centralization leads to the development of a series of major control centres within the late capitalist system, therefore. At the multinational level, these are the 'world cities'. Within a country, these are the centres of economic power. Some countries may have only one of these; in many cases it will be the original gateway primate or one of them in the case of a country with several such major centres. In a small number, most notably the United States, there may be several such centres, housing the organizing nexuses of a large and complex late capitalist economy. At all levels a characteristic of such systems is that most of the interaction is between the major centres. Flows of goods, capital, ideas and people are no longer dominantly up and down the local central place hierarchy. These continue, but are overshadowed in their volume by the flows between the major world and national cities, between London and

Paris, for example, New York, Chicago and Los Angeles, Toronto and Montreal, and Sydney, Melbourne and Tokyo. Late capitalism has introduced a major restructuring of the world's urban systems, in which much of the economic organization is colonially organized from but a few places.

Within late capitalist economies the major growth in employment has been in the white-collar occupations, including governments and their associated bureaucracies. Many of these replace, and expand on, jobs which were formerly in the small settlements which housed independent factory-firms and service establishments. Their concentration in a few large places deprives the smaller places of a certain class of residents, having both social and economic consequences for community structure. The seed-bed process is stunted, for example, as competition with large conglomerates creates very high barriers to entry for the person with a small volume of capital to invest: many parts of the urban system are, in effect, colonies.

As already suggested, the concentration of ownership and control functions, including the large sector which finances and services this sector, has been accompanied by the deconcentration, spatially, of the manufacturing functions, especially those involving routine and repetitive operations. This would seem to be part of the cost-saving activity of the firms, for not only is land cheaper in smaller settlements but also, and much more importantly, labour is too. In general, the larger the place, the higher the wage paid for similar work. In part this reflects differences in labour productivity, and in part the varying ratios between supply and demand in the relevant occupations, but analyses in several countries suggest that in addition the higher living costs in the larger places (comprising dearer housing, longer and more expensive commuting, and more negative environmental externalities, for example) are compensated for, usually after union pressure, in higher wage levels. Unions are generally easier to organize in large than in small factories, and so they are usually strongest in the large establishments of the big cities; this is certainly the case in the U S A. Firms seeking to reduce their labour costs may therefore decide both to transfer their manufacturing and other routine operations to smaller towns, where unionization is weak, wages relatively low and staff turnover rates less than in the larger places, and to subdivide their operations into smaller units, foregoing some possible economies of scale for operational convenience. With cheap and rapid transport available, perhaps within the firm, for high-value, low-bulk goods, the extra costs incurred in movement are more than offset in the labour and other cost savings from deconcentration.

Thus in the United States many companies are now establishing factories in the South, while continuing to organize their operations from the main centres of the north-east.

The need to cut labour costs has been particularly acute during the crisis of late capitalism which has characterized the 1970s, with high levels of both inflation and unemployment and low rates of economic growth. Many firms are producing below their capacity, and providing poor returns for their investors. To improve their profitability in the face of (at best) slowly growing markets, they have sought advanced technology which will allow them to employ mainly unskilled labour – often female, which is usually less unionized; only a few highly skilled employees have been retained to maintain the machinery. Thus the deconcentration to the small towns and their pools of cheap, non-unionized labour is part of the late capitalist dynamic. Indeed, some multinational firms have extended it even further by moving the components of their operations requiring unskilled and semi-skilled workers, supervised by a few outsiders, to countries where labour is very cheap, such as Taiwan, Korea and Mexico. The parts manufactured in these countries are sent to the neo-colonial centre for assembly into the final product. Other countries invert the process, by importing cheap labour from areas of high unemployment and under-employment, continuing the immigration policies practised by New World countries in previous decades, but refusing to grant these workers citizenship in case a depression in the economy requires that they be repatriated rather than become a burden on the social security systems of their hosts.

The tertiary and quaternary sectors

One of the major impacts of late capitalism on employment structures has been the decline of manual work consequent upon the technological innovations associated with automation, and the rapid increase in the number and relative importance of white-collar occupations. Many of these are not in what has commonly been termed the tertiary sector, which includes retailing and other trading functions, but in the relatively new, and certainly rapidly growing, quaternary sector, comprising the infrastructure for trade and finance, the state bureaucracy, many service establishments and functions, and education and other social services. Indeed, without this expansion late capitalism, because of its drive to achieve ever-greater productivity in manufacturing and agriculture, would have been characterized by very high levels of unemployment.

Some of this growth in the quaternary sector is the direct consequence of concentration and centralization, as described above. The large firms themselves have extensive employment in routine administrative tasks, as well as a considerable hierarchy of decision-makers, but most of the extra employment is outside the large conglomerates. A wide range of services – many of them either legal or financial – is needed to articulate the movements of capital and goods within a late capitalist system, and to link its components both with each other and with their major allies, the state systems. These now dominate the employment structures of most late capitalist countries, and certainly of their major urban settlements. Governments, for example, play a major part in the reproduction of the labour force, through the provision of education and health services, plus either a large component of, or a considerable stimulus to, the housing market. Certain aspects of these functions are decanted to local governments, providing the services where they are in demand, but their control, plus others which are closely tied up with support for capitalist enterprises, must of necessity be in the larger places, close to the centres of economic decision-making.

Other tertiary and quaternary services which have expanded under, if not been invented by, late capitalism involve those concerned with recreation, leisure and entertainment. The overall general prosperity; the shorter working day and week; the longer holidays; the earlier retirement and the longer life expectancy: all of these combine to generate demand for a wide range of passive and active pursuits. Some of the employment created to meet these demands is spread throughout the urban system, serving the demand where it exists. But some is highly centralized (spatially), particularly with the concentration and centralization processes which have occurred recently in, for example, the mass media and professional sport.

None of these trends involved in the concentration and centralization processes has ever been completed, however. Thus in manufacturing, for example, although large proportions of most industrial sectors are owned and controlled by just a few firms, there are still many viable small ones: complete monopolies are rare, except in small countries, and no capitalist country is likely to be controlled in the foreseeable future by a single firm. In part this is because it is not worth the while of the large conglomerates to take over certain firms or even certain sectors: many necessary processes involve only small production runs and require specialized facilities and personnel, and these are best provided by small firms. Such firms are dependent on the large ones for their markets, and thus for their survival, but their existence allows the conglomerates to direct their investment into potentially more profitable lines, while at the same time continuing the seed-bed process and encouraging enterprise among the would-be self-employed. The market that these small firms serve may be spatially defined; in restricted isolated areas investment by the conglomerates may be non-viable, so that environmental variations help to maintain the small-firm sector.

Late capitalist economies can be divided into their central and peripheral components, therefore; for the latter, 'peripheral' often refers both to their position in the size distribution of firms and to their spatial location. The central firms are the economy's leaders, on whom the small ones depend.

The urban system summarized

Industrial capitalism inherited a settlement pattern comprising three main elements: a low proportion of the population living in urban places countered by a large proportion of the workforce employed in agricultural pursuits; a hierarchy of central places through which trade was articulated; and, in many areas, a dominant gateway primate city. As industrialism proceeded, the first was changed substantially, so that by late capitalism most of the population lived in urban places. The first two have not been altered to any great extent, however, so that the main lineaments of the settlement pattern have remained the same, with variations in emphasis on different-sized places.

Capitalist success has been built on increases in labour productivity, initially in agriculture and then in manufacturing too. Thus capitalism has been dependent on the countryside, but its major impress has been in the towns: industrial capitalism is very closely associated with urbanization. Within the urban systems it has had three main impacts. The first has been the addition of factories and shops, following the general rule that the larger the place the greater the number of both that have been opened (and survived) there. Secondly, these establishments have generated further growth, through the multiplier process. And thirdly, it led to the creation of new towns, many of them based on locational advantages with respect to certain resources. With the exception of the last of these, therefore, urbanization has cemented the pre-existing settlement framework and ensured the continued relative importance of the mercantile towns and cities.

Late capitalism has succeeded industrial capitalism, introducing a new size dimension to the organizational characteristics of production, distribution and exchange. Again, this has not altered the existing settlement pattern significantly but has rather led to various modifications. The main 'change' has been the relative, if not absolute, decline of many of the smallest settlements coupled with the growing size and hegemony – political, social and economic – of the largest centres. During this redistribution, many of the new towns of the industrial capitalist era have declined along with the small central places, so that of the two main components of mercantile urban systems – central place hierarchies and large-city primacy – late capitalism has in general enhanced the latter rather than the former. Societies are now articulated through a few large cities only, both within countries and internationally.

The trends and processes described in this chapter have been predicated on a wide range of technological developments, some of the most important of which have been concerned with transport and communications. In late capitalism the latter have led to some reversal of the main trends concerning urban systems. Spatial concentration and centralization have been accompanied, on a more restricted, intra-regional scale, by deconcentration and decentralization. Urban sprawl and the creation of extensive urbanized regions, such as the Megalopolis identified in the north-eastern United States which extends from Boston to Baltimore, are typical of late capitalism, although it is more controlled in some countries than in others. These areas are far from entirely built over, but they are all part of the single functional unity. They are part of the single major trend of the industrial and late capitalist dynamics: the big get bigger.

5. Contemporary Urbanization, Urban Patterns and Urban Problems

The evolution of urban systems, as their encompassing societies have passed through the different types of socio-economic organization, has been described in the previous two chapters. Not all systems carry the imprint of every type of organization, but the major contemporary characteristic of almost all is their dominance by one or a few very large urban settlements. This dominance is variously viewed by commentators as either an advantage or a problem: the present chapter discusses the elements of the debate between the competing views.

Almost all modern urban systems have had their major lineaments determined by the forces of either mercantilism or capitalism. These forces involve competitive processes, and in all competition there must be losers as well as winners. The competitors are firms, households, communities and societies. All of these occupy locations, however, so that an urban system may be characterized by a clear distinction between the winners and the losers in terms of where they are located. Losing can bring problems to places with concentrations of unfortunate sufferers; winning can bring difficulties too.

The whole development of urban systems involves the spatial manifestation of winning and losing, of course, as was detailed in the preceding chapter, and this manifestation has in the past been accepted as part of the competitive process. Often counterbalancing forces have been set up within the economic system which prevent the winning places becoming totally dominant. Thus if a system contained two major urban places, X and Y, for example, and whereas X had much unemployment Y had many unfilled job vacancies, several consequences might follow: (1) the different supply:demand ratios for labour in the two places would mean that wages would be much higher in Y (where there were more vacancies than applicants, and workers would have to be tempted by higher wages) than in X (where there were more unemployed than vacancies, and wage levels could be kept down); (2) the unemployed would move from X to Y to find work; (3) employers would set up factories in X (perhaps moving their operations from Y) to benefit from the cheap labour supply available; and (4) these migrations would alter the supply: demand ratios and lead to an equalization of wages between X and Y. If both labour and capital were completely mobile, eventually an equilibrium might be achieved, with wages in X and Y being equal. Further changes – a factory closing in one place, or another expanding in the second – would create further disequilibrium. The pace of change, and the absence of complete mobility, will almost certainly ensure that no lasting equilibrium is ever achieved, but the operation of capitalist economics by both employers and employees should, according to this argument, ensure that wage differentials between places are not great, especially over a long time period.

Increasingly, however, societies have been unprepared to accept even temporary disequilibria, for a variety of reasons. One is that once a major disequilibrium develops, it is as likely to result in increased differentials as it is to generate counterbalancing forces. Decline in a major industry in a place may initiate a downward spiral, via the multiplier, which it is extremely difficult to reverse (as was discovered in most industrial capitalist countries during the 1930s). Another is that under late capitalism in particular, the relative strengths of the capitalist and working classes have altered somewhat, in favour of the latter, consequent on such trends as increased unionization and the extension of the franchise. The working classes have demanded state action to counter recessions and to protect their members from the vicissitudes of disequilibria; the capitalist class has demanded similar protection, pointing out that, because of concentration and centralization, recession in but a few firms can seriously harm employment and income prospects for large sectors of the population. Thus to maintain full employment, to guarantee capitalist success (as far as is possible), to protect members of the working class who might suffer from various consequences of capitalist competition, and also to shore up their own electoral prospects, governments in most countries have become active participants in their local economies. Much of this activity involves various forms of economic and social planning; the form of relevance to the theme of this chapter is variously known as urban planning and regional planning, among other titles.

Approaches to urban and regional planning

In its broadest sense, planning incorporates any activity intended to realize a predetermined goal: it is a means towards an end. In urban and regional planning, the end is a desired spatial distribution – of people and of jobs, for example. Planners are employed by governments to advise on the definition of such ends, on the preparation of policies which should achieve those desiderata, on the monitoring of the plan in operation, and on any necessary modifications to the plan in the light of new information (including that obtained from the monitoring exercise).

The definition of ends or goals for an urban or a regional plan can be undertaken in a variety of ways. One geographer-cum-planner, Brian Berry, has suggested a classification comprising four types of planning strategy, and this will form the basis of the present discussion. The first is *ameliorative problem-solving*, which involves planners (or their political masters) perceiving a current problem and taking steps to identify its causes prior to development of a policy which will remove it. This is piecemeal surgery, therefore. Some aspect of the present urban system is identified as a problem, and so the system is restructured in an attempt to eradicate it. The competitive processes then operate in the new format, and almost certainly create further problems for amelioration, as consequences of the policies enacted to counter the first. Such problem-solving is a continual process; the solutions produced by the planners of one generation may turn out to be the problems for their successors (as with the high-rise blocks of flats built in many cities in the 1950s and 1960s). For most governments, the major regional and urban problem to which they respond is unemployment.

The next two approaches to planning focus on trends. The first, *allocative trend-modifying*, identifies current developments and predicts their likely influence on the urban system. The various predictions are scrutinized, and evaluated as to whether their likely end-states, at a given date, are desirable or undesirable, acceptable or unacceptable. Planning policies are then designed which will encourage the desirable and discourage the undesirable, thereby steering the system towards the best of the many possible states towards which it is heading. The other, *exploitative opportunity-seeking*, is somewhat less goal-oriented than the previous one. Having identified the current trends, planners and politicians decide which offers the best immediate prospects, in terms of their goals; they ignore the long term in order to obtain immediate gains (perhaps supported by a view that, in any case, the longer-term cannot be predicted with any certainty).

The fourth approach, *normative, goal-oriented planning*, is the most utopian, for it involves a clear definition of a desired future state of the urban system prior to any creation of policies. This approach is not a set of reactions to current problems and trends, therefore, but since virtually no urban and regional planning begins on an empty landscape, the format of the utopian design is undoubtedly strongly constrained by the morphology of the present system. (The life of most buildings is measured in decades, for example, and the capital costs of most transport facilities demands that they be used for similar periods.)

Of these four approaches, the last is rare in capitalist societies, despite lip service to it and much preliminary effort in the preparation of outline desired states (few justify the term utopian). Immediate gains are much more important to the various interest groups in such societies, for both capitalist and working classes. In the private sector this is because the success, and survival, of firms is dependent on a continued satisfactory profit level, so that the horizon with regard to investments and research (except, perhaps, for the largest of firms) is only years rather than decades distant. In the quasi-public sector (as with semi-independent nationalized industries) the profit motive is also important in many cases. Governments, too, focus almost entirely on short-term horizons, both because of their links with the capitalist sector and because their legitimation by the electorate is dependent on policies which bring current success and are thus popular with the voters: only a totalitarian government can survive for long on policies which offer blood, sweat and tears for today and tomorrow, but plenty on the day after.

Within capitalist societies with liberal-democratic governments, therefore, planning policies are likely to combine ameliorative problem-solving and exploitative opportunity-seeking. The following sections discuss some of these policies as they relate to perceived problems of and in the urban system. Some countries, however, have developed other forms of social, political and economic organization, and these are reflected in their planning policies. In almost every case their urban systems have been inherited from mercantile, if not industrial capitalist, forms of economic organization, and these have acted as constraints on the policies introduced. The final section of the chapter looks at some of these countries.

Urbanization and over-urbanization

According to earlier discussions, the rate of urbanization in a country – the redistribution of population from rural to urban – is closely associated with its level and rate of industrialization. Thus in the mercantile era the majority of the population lived and worked in the rural areas; the towns represented a small, though focal, element in the social and economic organization. As industrialization proceeded, so labour was released from rural occupations (often forcibly) with the drive for greater productivity and profitability in agriculture and the replacement of local craft workshops by urban factories. To find employment, these displaced rural workers found it necessary to move to the towns, and as industrial capitalism progressed so the proportion of the population remaining in the rural areas fell until, as in the United Kingdom, it is now less than 10 per cent. (One of the problems of quoting such figures concerns the definition of 'rural'. Increased mobility means that many workers in urban areas can live in the countryside, whereas others who lived and worked in the towns and cities retire to rural areas, where they have no direct contact with the local economy. Many countries now comprise a mosaic of overlapping urban regions, containing much of the rural area.)

This spatial redistribution of population implies a functional relationship between industrialization and urbanization, of the generalized form shown in Figure 5.1: the higher the level of industrialization (the per-

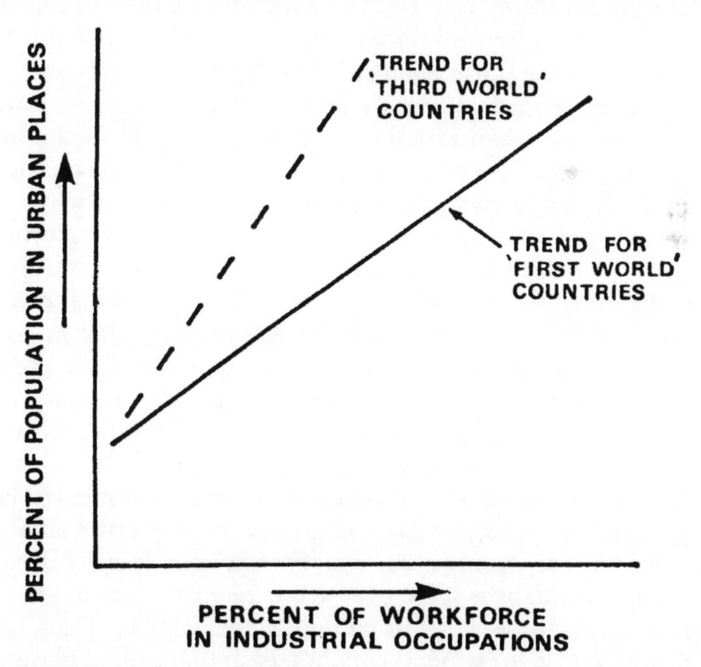

Figure 5.1 The relationship between industrialization and urbanization in the 'First World' and in the 'Third World', indicating the differences between the two that suggest that countries in the 'Third World' are over-urbanized relative to those in the 'First World'.

centage of the workforce in industrial occupations), the higher the percentage of the population living in urban places. Such a relationship can be traced in the case of each individual country, or can be generalized for a single temporal cross-section by using data for a selection of countries. When the latter procedure is employed, however, it is commonly noted that countries of the 'Third World' generally have a higher level of urbanization, relative to their level of industrialization, than do those of the 'First World' (the countries of Europe and North America which industrialized in the nineteenth century). This has suggested to many commentators that the countries of the Third World are over-urbanized, that for their level of economic development they have too great a proportion of their population resident in towns and cities.

The evidence on which the conclusions regarding over-urbanization are based is not overwhelming, particularly when the urbanization of individual Third World countries over time is compared to that of individual First World countries. Further, variations among the latter group of countries are greater than those among the former in the level of urbanization which can be associated with a particular level of industrialization; this leads to queries about the validity of setting the 'First World average' as the norm against which Third World experience is compared. Nevertheless, avoiding the emotive term 'over-urbanization', it does seem that the urbanization process has been somewhat different in these two groups of countries, and it is necessary to reconsider the analyses of the preceding two chapters in order to understand why.

Colonization occurred through much of the Third World during the mercantile period, as part of the search for cheap raw materials for the industries and consumer markets of the First World. The colonial land was exploited by the immigrant merchants and agriculturalists. The wages paid to indigenous workers were low, a portion of the profit on their production remained in the port cities, to pay the colonial administrators and military, and perhaps also to recompense the local élite, and the remainder of the surplus was retained by the colonial investors in the First World homeland. It was used to stimulate growth in the latter, via the multiplier process, and as a consequence was not available to generate economic, especially industrial, growth in its country of origin.

As industrial capitalism took over from mercantilism in the First World, three processes were set in motion in the colonies. First, efforts were made to increase the productivity of labour and capital in the agricultural sector, thereby both increasing the profits for the homeland investors and reducing the price that they had to pay for their materials. Second, the colonists introduced health services, often through the medium of the state, thereby externalizing the costs which might otherwise have to be met by the employers of the colonial labour force. This improved the quality of the indigenous labour: diseases were conquered, mortality rates (especially those of infants) were reduced substantially, and life expectancies were increased. But, as a counter to these changes, neither the colonial attitude to nor, later, their technology for birth control was as successfully introduced. Thus together the release of labour from the land and the rapid rate of population increase stimulated by the reduced mortality meant that there was a very large 'surplus' rural population for whom there was neither work nor prospects. As capital-intensive farming replaced the preceding labour-intensive activities, as peasant cultivators were dispossessed by the aggrandizement of large, export-oriented farms, and as population growth boomed, so the fabric of rural society was destroyed, and breadwinners and single individuals, if not whole families, were forced to move to the towns, of which there was often only one – the gateway primate.

These two processes operated also in the First World during the early decades of the industrialization process (and its associated 'agricultural revolution'). But the rapidity of change in the countryside, the volume of the migration stream, and the associated poverty and underemployment in the urban areas were all much less in the case of the First World. Despite the appalling conditions recorded in, for example, parts of London during the nineteenth century, in general the disequilibrium between rural release of population and its absorption in the urban areas was less there then than it has been during the twentieth century in the Third World. The reason for this difference lies in the third process.

As industrialization proceeded in the First World, so problems of overproduction began to arise in certain industries. Technological improvements allowed increases in productivity to outpace the expansion of home markets, and so the Third World countries had to be exploited as customers for First World industries. Merchants and industrialists in the latter countries combined to ensure that colonial administrators, in the 'homeland' as well as in the colonies, limited local colonial manufacturing development to products which could not readily be imported. Thus not only was much of the surplus (the difference between 'making price' and 'selling price') produced by colonial agriculture removed to the homeland but also much of what remained in the colony was, as far as possible, to be spent on imported goods (and services, such as shipping): the multiplier was to be kept in the homeland.

As decolonization proceeded, so administrative fiat preventing the establishment of local manufacturing industries in the newly independent states disappeared. Indeed, the new governments were keen to encourage industrialization, to absorb labour, to stimulate growth and to reduce dependence on outside suppliers. This was countered by the First World through its great strength in economic competition, involving such tactics as price-cutting (as described on p. 112) as a means of defeating local competition. Protectionist policies may be used to combat the price differential enjoyed by the First World producers, but these can easily mean that local production is expensive, and the multiplier sluggish. Further, to foster local industries Third World governments have often had either to borrow capital from the First World, at terms which ensure that some of the surplus is expatriated (this is also true of many inter-governmental aid programmes) or to invite overseas firms to establish plants in their countries. The latter have often received subsidies (later expatriated as profits) and the factories which they have built are usually capital-intensive and so absorb little of the large pool of labour available. (The exception to this is the movement of labour-intensive manufacturing processes to Third World countries by multinational firms, but these, too, create little in the way of a local multiplier.) However a Third World country attempts to industrialize, therefore, in most cases a sizeable proportion of the surplus value has been exported to the First World, and the local search for growth and further employment has been retarded.

In the Third World, therefore, the demographic base to urbanization has been different from that existing in the First World during its process of industrialization, and the economic base has been so structured by the forces of colonialism and neo-colonialism that a considerable portion of the stimulus to growth has been externalized, creating jobs and wealth in the First rather than in the Third World. Cities in the latter are very large, and their process of economic development has produced a greater degree of concentration in a smaller number of places than was true in much of Europe, thus making the problems of urban unemployment and poverty even more obvious. The consequence is the apparent problem of over-urbanization.

Some countries outside the Western European heartland of industrialization have avoided many of the problems of economic dependency after the withdrawal of the colonial administration. Most notable among these have been the settler colonies of the temperate lands, where immigrants from the colonial countries were granted permanent property rights and so provided with the basis for producing a local surplus that could be invested for local gain. (Most of these countries borrowed heavily from the First World, of course, and their ability to generate much growth through an indigenous multiplier depended on an ability to sell agricultural and other products to their former rulers at reasonable prices.) The larger the country, the greater the benefit obtained in general, so that the main example of this type of country is the United States; Canada is smaller, and has suffered recently from American neo-colonialism, whereas both Australia and New Zealand – plus the rather special case of South Africa – have been disadvantaged by their small markets (both have been neo-colonized by Japan to some extent). Other countries, with even fewer settlers prepared to invest in their new home, have been slower to develop indigenous industries – Brazil, for example. Only Japan stands out, among capitalist nations, as having generated industrialization and urbanization indigenously, with relatively little outside participation.

Over-urbanization as defined above, therefore, reflects the type of economic and social organization imposed on much of the present Third World by colonial political masters, and the inability of governments and capitalists in these Third World countries to counter the economic power of their former rulers. The consequence is not only a high level of unemployment in the cities of the Third World, and an inability of the state to provide a bulwark against the resulting poverty, but also a very high level of underemployment. Jobs in manufacturing industries are few, and a large proportion of the urban population is forced to work in the tertiary and quaternary sectors where productivity and wages are in general very low. In retailing, for example, there is a large 'informal' sector of sellers in most cities, working in the 'bazaar' or indigenous economy; service industries proliferate in this sector, with a large, labour-intensive, self-employed component. And in some countries, notably in Latin America, there is much employment of cheap domestic servants. In these ways, what proportion of the surplus from capitalist farming and industry is captured locally is spread very thinly and circulates relatively slowly; its multiplier impact is often slight.

Over-urbanization, as it is commonly understood, is thus a reflection of the international economic, social and political structure. The problems which it embodies are only problems which involve potential solutions within the fields of urban and regional planning to the extent that the cities are too large. This question of the 'problem' of urban size is a general one, and certainly has not been addressed in the Third World alone.

The problem of urban size

The problem of over-urbanization (if it is perceived as such) is largely confined to the countries of what is known as the Third World. Over-urbanization is usually associated with very large cities, however, even in those countries lacking a dominant primate gateway centre. But big cities characterize nearly all countries, in most of which their existence is seen by some analysts, commentators, politicians and planners as a major problem. For some the problem statement reflects a utopian view of rural and small settlement life, a view which cannot be answered by analytical argument. For most, however, the perceived problems of large cities are tangible, if difficult to measure.

There are, of course, those who argue that urban size is not a problem in capitalist countries, for if it was the forces of capitalism would react to it and produce an alternative, and more efficient, population distribution. Indeed, they point out that such a reaction is currently taking place, with the decanting of certain types of industrial and administrative operations from the big cities to smaller places. Their opponents point out that this trend is a relatively minor one, and is not challenging the hegemony of the big cities. In any case, they contend, *laissez-faire* analyses are brought to a conclusion that big cities are not a problem because their definition of efficiency is that which exists.

Urbanization offers benefits to firms: large places provide access to markets, to linked industries and to pools of trained labour, and capitalizing on this allows firms to reap the economies of scale, both internal and external. The result of such benefits should be reflected in profit rates, so that the larger the place, and the greater the economies that it offers to firms, the higher the profit levels that should be achieved. This is suggested in Figure 5.2A, which illustrates what may be the relationship between city size and marginal benefits, the returns to a firm from expanding its production by one unit. (The graph also indicates the benefits to all firms from growth of the city.) Of course, firms meet costs as well, and so a hypothetical marginal cost curve is also shown in Figure 5.2A. For the smallest cities, marginal costs of production exceed marginal benefits, which indicates that firms there would make a loss. Costs fall with increasing city size, up to a particular level: at a certain size marginal benefits exceed marginal costs, and this can be assumed to indicate the minimum size of settlement for viable firms. Beyond that level, the larger the city, the greater the gap between costs and benefits (despite an increase in marginal costs above another size threshold,

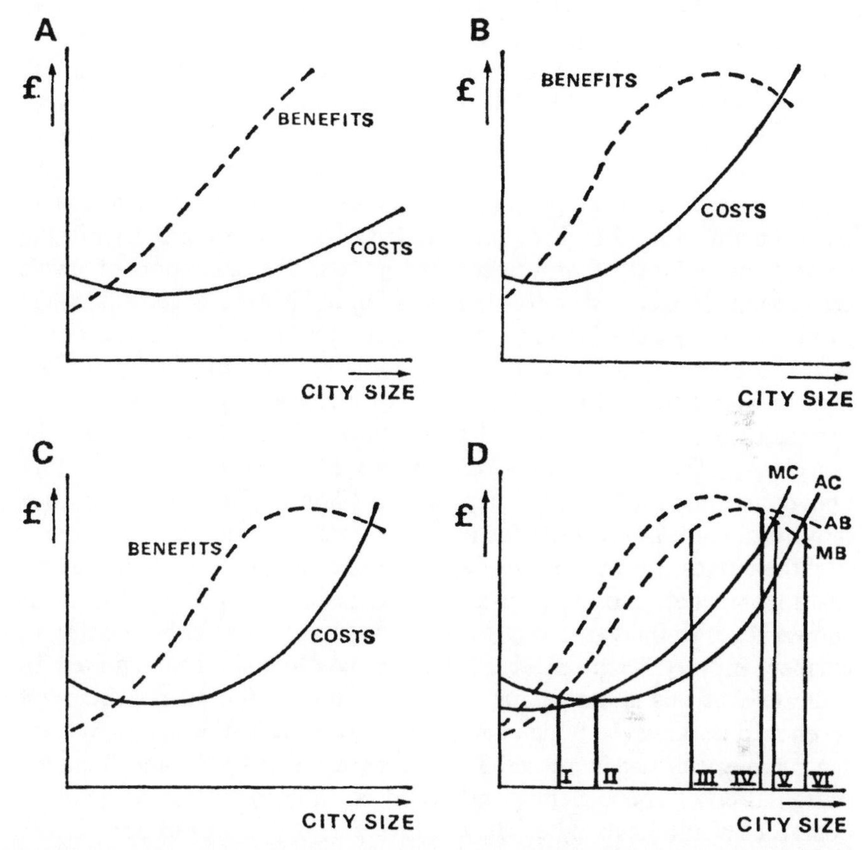

Figure 5.2 The costs and benefits of increases in city size. A and B show the marginal costs and benefits associated with city growth; C shows the average costs and benefits; D shows both marginal and average trends (AC=average costs; AB=average benefits; MC=marginal costs; MB=marginal benefits). The curves in B, C and D are based on Richardson, 1973, p. 11.

reflecting such 'problems' as traffic congestion), and so the greater the profitability of firms. Thus, because of multiplier effects and the relationship of size to the economies of scale, in industrial terms it would follow that the bigger the city the more efficient it is as a locale for productive activities. And, it is claimed, because an efficient industrial system is the basis for an efficient society, if big cities are in the best interests of firms then they are in the best interests of everybody.

The counter-argument to that based on Figure 5.2A is that the gap between benefits and costs does not widen continuously as cities grow. Beyond a certain size, it is contended, the rate of increase in benefits may slacken, whereas that of costs may grow. Internal economies of scale may not increase beyond a certain size level: more production and greater productivity on the shop floor may be countered by increased administrative costs involved in linking the many specialized departments of a very large firm. Similarly, external economies may turn into diseconomies above a certain size, since all the links a firm needs are now present. The larger the place, the longer the average length of journey involved in linking firms with other firms and with customers, for example, while the greater the volume of movement, the greater the likelihood of traffic congestion: longer and slower journeys impose greater costs. And wage rates may increase, relative to productivity, because of supply: demand ratios and unionization levels; environmental pollution may impose extra costs; social disorganization may increase and involve greater expenditure in its control and alleviation: all of these increase the marginal costs of production. Thus the curves may approach each other above a certain city size, as suggested in Figure 5.2B. Eventually they may meet, at the point where further growth may be counterproductive.

If the curves shown in Figure 5.2B are accurate reflections of the true relationship between size, costs and benefits, then presumably firms should react to increased city size beyond a certain level by moving to smaller, more efficient places. Evidence has already been quoted to indicate that this is happening in some countries and cities. There are mechanisms, however, which delay the realization that diseconomies are setting in as city size increases. The curves in Figures 5.2A and B are for marginal costs and benefits, and could be interpreted as showing the costs that would have to be met, and the benefits that would be received, with each increment of city growth. Thus with costs, for example, it may be that in a city with 100 factories the total cost of production is £1,000,000; with growth to 101 factories it is £1,011,000, to 102 it is £1,022,500, and so on. The first additional factory increases costs by £11,000 and the next by £11,500. But many of these costs are externalized, because they are services consumed by all (some of them provided as part of the collective reproduction by the public sector), for which an average price is paid. Thus the average costs per factory (each of which is assumed to be producing the same volume of the same commodity for this analysis) increase from £10,000 to £10,009·9 with the introduction of the 101st factory, and to £10,024·5 with the 102nd. The impact of growth is being spread among all, therefore, and the firms which generate growth do not pay the marginal costs. Average cost and benefit curves approach each other more slowly than do the marginal curves on which they are based, as suggested in Figure 5.2C.

If these average and marginal curves are superimposed (Figure 5.2D), several critical city sizes can be identified. The first (size I) is the minimum viable size; below this the costs to a firm exceed its returns per unit of production. Size II is the least-cost size, at which production is cheapest; it is much smaller than that at which returns are greatest (size IV). At size III the gap between average costs and average benefits is at its widest point, so this is the most profitable size. Size V represents the point at which marginal costs come to exceed marginal benefits, and is the one above which the city should not grow since additional production is not profitable. Firms may not realize this, however, because their calculations and their taxation rates are based on average costs; profitability continues until size VI is reached. Assuming that these curves in Figure 5.2D are reasonable representations of the as yet unmeasured costs and benefits of growth, then proponents of a *laissez-faire* approach to urban size would argue that beyond III firms will begin to move away from big cities, and that there would certainly be no growth beyond VI: firms may not realize that growth beyond V is undesirable, however, so that planning policies may be required to ensure that expansion beyond that point does not occur. Presumably, therefore, as most large cities in the world are continuing to grow, size VI has not been reached, although III may have been in some cases. (The horizontal scale for Figure 5.2 may not be linear in the 'real world', of course.)

One problem of such an analysis, however, relates to the question 'Who benefits and who pays?' As indicated earlier, firms have externalized as many of their costs as possible in late capitalism by requiring that the state pay for many aspects of the reproduction of the labour force and of the infrastructure for successful production. In this way, some of the average costs of production are met by society as a whole, so that the employees of industrial firms meet part of the costs of those firms, and thus 'subsidize' profits out of their wage packets. Thus if capital can internalize its benefits (pay as little as possible in wages and taxes) and externalize its costs (cleaning up its pollution, for example), it can go on benefiting after sizes V and VI have been reached on Figure 5.2D. (In effect, it pushes down the average cost curve.) Further, the greater the concentration in an economy, the better able are firms to pass on increases in costs to their customers and maintain their profit levels (thereby, for example, externalizing wage increases and encouraging inflationary wage demands).

Although large cities may be relatively more costly than smaller ones, once the former exist the potential for redistributing population and creating an alternative city-size distribution is usually slight. As indicated earlier, the hegemony of large-city firms allows them to undercut small-firm rivals in any price competition, and thereby presents a major barrier to the establishment of new firms in smaller places. (Pricing is not the only tool used in competition. Large firms, especially conglomerates operating in several sectors, can 'tie' customers by refusing to sell them certain needed commodities if they buy others from their competitors.) And since most of the small firms on the periphery of the economy are closely tied to the large conglomerates, any uncertainty which their owners may feel about location decisions will lead them to take the conservative route and cluster in the big cities close to their major links.

The links between firms, and the existence of highly interdependent industrial complexes, provide another reason for the continuation of large cities beyond even size III in Figure 5.2D. It may be that an entire complex could profitably move to a smaller place, but no one individual firm can afford to pioneer the move. This situation is known in mathematical game theory as the 'prisoner's dilemma', and can be illustrated by the following simple example. The data in the table show the profit rate (in percentage terms) for firm A and for all the other firms in an industrial complex of which A is a part, according to the size of town in which they are located. In each part of the table, the first figure is the return for A and the second is the return for all others.

		All Other Firms	
		Small Town	*Large Town*
Firm A	*Small Town*	8,8	2,10
	Large Town		6, 6

The bottom right-hand cell shows that in the large town all firms achieve a profit rate of 6 per cent. If A was to move to the small town, and the other firms did not, A would earn only a 2 per cent return whereas the others would get 10 per cent. (The unlikely event of all the firms except A moving to the small town is not indicated in the table.) If all firms were to move, each would get a return of 8 per cent. A move would be beneficial to all, but no single firm, such as A, could take the initiative and gamble that the others would follow. This suggests that some external planning agency, the state, should indicate to the firms in the complex the benefits of a move, and thus protect any would-be gambler. Many planning agencies have attempted such strategies, but have largely failed, in part because industrial complexes are rarely independent units (if they were, they would probably be incorporated into single firms under the processes of concentration and centralization) and enough members of any one complex are tied to other complexes as well to militate against their movement (unless all complexes move!). And in addition most firms in large cities have major investments there, for which they would require substantial compensation in return for agreeing to move.

Together, the prisoner's dilemma analogy and the various issues involved in the accounting procedures used in assessing costs and benefits suggest that individual firms are unlikely to take the major industrial location decisions that will substantially influence the size distribution in an urban system. Some are moving certain functions away from the big cities (p. 117), but are countering this trend by the expansion of the functions remaining in the large places; the latter trend is accompanied by expansion of the linked large-city functions, notably in government activities. Only society as a whole, through its public policies, can achieve spatial decentralization and a reduction in the dominance of large cities, therefore. This has been realized for several decades by planners and others, and emphasis was given to the case by the development of aerial warfare during 1939–45. Thus in France, for example, much effort has been expended in recent decades on halting the rapid economic growth of the Paris region and on redirecting activities to peripheral metropolitan growth points. In Britain, similarly, attempts have been made to weaken London's hegemony and either to direct or to encourage firms to move to other centres. Evidence of the last decade suggests partial success for such policies, though some argue that the amount of decentralization achieved would have occurred without government expenditure as capitalists sought cheaper labour. The British government has moved many of its own offices performing routine functions which do not have to be carried out in London close to the heart of government, although such moves have not been popular with the civil servants involved. Except in cases when governments have refused to allow further industrial development in certain places, thus ensuring growth elsewhere, state activity aimed at a redistribution of industry and population away from the large cities has largely involved the use of incentives, such as cheap loans and factory premises. (Unfortunately, it is almost impossible to ensure that firms meet the marginal costs of their location decisions which, as Figure 5.2 indicates, might encourage movement at a smaller size.)

The subdivision of the state machinery itself in most countries into national and local governments frequently hinders the operation of any incentives policy. Local governments provide much of the infrastructure for economic and social activity and are the institutions to which many business costs are externalized. In many countries a major source of revenue for these local governments is a tax on either local property values or local incomes. Both are dependent on local prosperity, so that without growth in its territory, and thus an increasing tax base, the local government's ability to raise revenue declines, and either its level of service provision must deteriorate or its provision becomes more expensive. Thus local governments cannot afford not to grow, because this interferes with their ability to provide necessary services to those already present and increases average costs for residents (relative to those in growing places). In particular, an absence of growth and a declining tax base will make it very difficult for a local government to renew the social and economic infrastructure which it provides, as illustrated in Britain in the 1970s when the government found it necessary to halt the new town programme in order to revive the rapidly declining inner-city areas.

In a late capitalist society, therefore, the conflict between and within the capitalist and working classes which characterizes industrial capitalism is joined by conflict between governments. This can be seen at a variety of levels. Within Australia, for example, a state's government may wish to halt the growth of the primate city and direct new industries to 'country towns', but if a multinational firm inquires about locations in the state and is told it can only go to a small town it may decide to go to another state instead. Thus the state government is faced with growth in its main city or no growth at all. The same is true between countries in a common market such as the EEC. Although decentralization and a reduction of the size of large cities may seem a desirable planning goal, therefore, it is often politically inexpedient: continuation of big-city growth rather than accepting the consequences of decline is a clear example of planning as exploitative opportunity-seeking.

The arguments for big cities, therefore, are that they are the most efficient for the operation of capitalist economies, otherwise they would not exist. Because they provide a variety of scale economies, they are more productive locales than are smaller places, and as a result their inhabitants are more prosperous. Furthermore, because they are the places where most patents are filed, they are clearly the most inventive centres, and thus the major engines of economic growth. Selective decentralization is occurring, where it is in capitalist, and therefore societal interest, but any further attempt to redistribute population and economic activity would be counter-productive.

The argument against big cities is that they are not necessarily as efficient as is claimed, in part because individual firms cannot initiate the sorts of moves which would increase efficiency if they knew that others were going to follow them, and in part because capitalists can externalize many of their costs, so that society as a whole must pay the bill for their inefficiency. The argument regarding productivity is answered by reference to the law of diminishing returns and by pointing out that a correlation (between city size and inventiveness) does not imply cause and effect; if necessity is the mother of invention, then capitalists will invent whenever they need to, wherever they are, and large cities, especially in an age of rapid communications, have not been proved necessary for this. (Much research and development work now takes place outside big cities.) Regarding higher incomes, arguments are advanced that these to a considerable extent reflect higher living costs in larger places (because of longer commuting trips, higher land values, etc.), and do not necessarily indicate that people doing a certain job in a large city have higher real incomes than their exact counterparts in smaller places.

Finally, the two sides in this debate over city size differ over the meaning of efficiency and growth in a capitalist economy. The usual definition of economic growth is an increase in Gross National Product (GNP) per person. GNP is the sum of all expenditure, so that if it increases more is being spent and this indicates growth and progress. But the building of an inner-city motorway may not be growth, it is argued, if this reflects the congestion created by large cities, and such expenditure could be avoided if big cities were replaced by a larger number of small ones. In other words, decentralization might free more capital for investment in facilities which would improve the quality of life rather than merely clean up the mess already made. As has been illustrated previously, all expenditure generates more through multiplier processes, and an accounts system which identified whether spending was to create growth or to clean up the problems of earlier growth might refute arguments that progress was taking place by showing that 'real' GNP was declining. For example, it may well be that the larger the city, the greater the accident rate for motor vehicles. Such accidents have two sets of effects. First, the injury to persons involves the ambulance service, doctors and other hospital staff and results in loss of production by those injured; in addition, the demands of the latter require the training of more medical staff to treat their injuries, and so on. Secondly, the damage to vehicles creates work for those who repair them, for the insurance assessors who work on the claim to cover the costs of repair, and for the manufacturers of components; it also leads to an increase in insurance premiums (externalizing costs again). All of these costs are caused by the accidents and if these could be avoided, or at least reduced, through a redistribution of population, it would be possible to replace expenditure on them by investment in 'goods' rather than the correction of 'bads'. Of course, an urban system comprising more relatively small places and fewer large ones will probably be more costly on certain grounds – in its use of fossil fuel energy for moving people and goods, perhaps – but the case against the current use of GNP statistics is that separation out of 'welfare' expenditures (those that improve societal wealth because they contribute to movement towards a normative goal) from 'illfare' expenditures (those involved in ameliorative problem-solving) could provide a useful basis for the planning of settlement size distributions.

Most of the debate about the apparent 'goods' and 'bads' of big cities focuses on those places alone, and does not set them in the context of the entire urban system which they dominate. A more holistic approach, it is argued, would indicate that the concentration and centralization trends which characterize the developing hegemony of the big cities have important negative consequences for those who remain in the smaller places. The latter, it is argued, suffer from a lack of amenities, especially social ones, from a limited range of job opportunities (including access to senior positions which are almost all available in the large cities only) and from the high prices which result from having to 'import' almost all that they consume. (In effect, small towns are similar to the small countries of neo-colonialism.) On the other hand, it is argued, they have pleasanter environments in which to live, and residents can buy homes much more cheaply, so that it is extremely difficult to draw up a balance sheet which shows who is 'better off', the big-city dweller or his small-town counterpart. In any case, some argue, almost all societies offer sufficient choice so that people can opt for the sort of environment they prefer; but even if this were the case for all, the costs of moving from a small town to a large one, and paying the much higher property prices, make exercise of that choice expensive for some.

The debate about the pros and cons of big cities is unlikely to be resolved, although some would claim that the coming micro-processor revolution will make it obsolete since proximity will soon be an inferior good. Big cities have both attackers and defenders. But big cities exist. Planners and politicians are more likely to act as ameliorative problem-solvers and exploitative opportunity-seekers with regard to the *status quo* than to introduce normative policies, whose impact can only be very long-term, aimed at rewriting the settlement pattern on a major scale. The big city may or may not be the most efficient locale for capitalist enterprise (the accounts still have to be drawn up), but while capitalist firms believe that it is, and act accordingly, it is unlikely that any planning activity will generate a major shift.

Urban sprawl

A major feature of most big cities is not their population size but rather the extent of their built-up area. In recent decades, as a reflection of increased personal mobility consequent on widespread car ownership, growth of the built-up area, in percentage terms, has been more rapid than that of the population. Cities have sprawled extensively, and now comprise regions of discontinuous urban development. Megalopolis, the urbanized area of north-eastern United States extending from Boston to Baltimore but including much non-urban land use, is often cited as an example of this, although other cities, such as Los Angeles, Johannesburg and Melbourne, have much lower overall densities. Further megalopolises are believed to be in the making, such as Chipitts (Chicago to Pittsburgh), Sansan (San Diego to San Francisco) and Boswash (Boston to Washington) in the United States and the Randstad in Holland.

Sprawl is widely believed to be undesirable. For example, it is often argued that it increases the costs of operating the urban place, not only for the firms which must spend more on transport and for the commuters who must undertake longer, more expensive journeys to work but also for the governments and agencies who provide basic utilities such as electricity, water and gas, sewerage and garbage removal, streets and roads. The lower the density of housing, the greater the cost per unit of providing all of these services and also the more likely it will be that public transport cannot be operated successfully, because of the long distances that must be traversed in order to fill a vehicle. Sprawl, it is claimed, is expensive, for both individual and society.

Sprawl usually involves the urbanization of what was formerly agricultural land. Many towns and cities, especially those of any age which were initially largely dependent on their immediate hinterland for food supplies, are in areas of fertile soils; indeed, in many cases the more fertile the soils the more prosperous the town and the greater its propensity to grow and sprawl onto the fertile land. Sprawl involves the loss of such good land to agriculture, since residential developers are usually able to offer more for land than it is worth under cultivation. Thus to many observers sprawl sterilizes a valuable resource, whose importance may increase rapidly if food shortages threaten. To others, however, increases in productivity on land not incorporated into built-up areas more than compensate for the losses involved in urbanization, and in some cases production from gardens exceeds what farmers previously obtained from the same land.

A final argument against sprawl is that it encourages financial speculation in land and results in its under-use. As a town grows, so the land around it becomes attractive to investors, who may buy it in the hope of reselling it at a high price for residential development. In the meantime, they may either let it lie idle or decline to invest in its fertility and so obtain maximum yields from it. (Alternatively, people may buy large plots of land on the edge of a city as 'hobby farms', again not working to obtain maximum potential yields.) Such investments take capital out of the multiplier process which stimulates industrial growth and the generation of employment, and so speculation harms the economy in two ways.

In some countries the perception of the problems of sprawl has led to the development of planning legislation aimed at limiting, if not halting, their impact. The planning process introduced in Britain after the Second World War, for example, included the provision of green belts around major cities which would limit their expansion: any growth which could not be catered for within the boundaries so defined was directed to communities beyond the green belts, including the new towns established in a ring around London. The latter were supposed to be self-contained, but the eight new towns now house substantial numbers of commuters who travel daily to London. Similar development of housing by private firms in towns and villages beyond the green belt has resulted in the creation of an urban region comprising functionally linked but spatially separated communities, a trend which has been countered by later planning with the establishment of larger new towns further from London, and thus beyond commuting range (for all but the most determined). Other countries have followed the British lead, in restraining sprawl if not in building new towns too, but in the United States, reflecting a different attitude to a commodity, land, which is believed to be in plentiful supply, and under pressure from the interest groups (such as the construction industry) which benefit from sprawl, there has been much less legislation aimed at inhibiting the expansion of built-up areas.

Planning legislation may reduce many of the impacts of sprawl and protect good agricultural land, but in many cases it does not reduce speculative activity. Thus, for example, the creation of a green belt makes the transfer of land it surrounds to eventual urban use almost certain. Furthermore, the existence of the limit to expansion fixes the supply of available land so that, assuming a continued demand let alone the probability that an increasingly affluent population will be wanting more space and being encouraged to aspire to new suburban homes, the supply:demand ratio for land will be changed. The result may well be a speculator's dream, a guaranteed artificial shortage of land bringing with it rapidly increasing prices. And developers may have to respond to this induced shortage by accumulating a stock of land sufficient for the next two years or more of building activity, without which they cannot plan their programmes. (The slowness of the machinery used to give planning permission for development may also encourage the growth of such land-banks.) The costs of holding land for this length of time (raising loans and paying interest) will have to be passed on to their customers, while at the same time the existence of the land-banks will accentuate the shortage of land for sale. Thus ameliorative problem-solving planning legislation aimed at limiting urban sprawl will result in a more compact, cheaper-to-operate city which is not too voracious in its consumption of formerly agricultural land. But without either supporting legislation to limit speculation or state activity in the land market (such as the municipalization of land around Stockholm) it may create another set of problems.

Counter-urbanization

In North America, western Europe and a few other late capitalist societies, recent trends suggest an accentuation of the issues related to urban sprawl. Until the 1970s, in most of those countries the large cities were net gainers from migration processes: more people were moving to the cities than away from them, although within the cities there was a deconcentration of people and jobs into the expanding suburbs. But then a major change occurred, over a short time, and the cities became net losers from the migration process. More people were moving to the medium-sized urban areas than to the largest. This process became known as *counter-urbanization*, and its extent was made clear by census results at the end of the 1970s. Between 1960 and 1970, for example, the population of metropolitan areas in the United States increased by 17.1 per cent, whereas in the rest of the country the increase was 10.0 per cent; in the following decade the respective figures were 3.9 and 15.4 per cent. Exactly the same happened in England and Wales: between 1971 and 1981 many of the main urban growth centres were in East Anglia.

This counter-urbanization trend suggests that the problems of the big cities discussed above have been realized, and that natural economic forces are producing spatial deconcentration – which is of jobs as well as people. Closer analysis has suggested that the shift of jobs reflects industrial restructuring in a period of economic crisis. Falling profits and markets mean that to survive firms must increase productivity. By closing their plants in large cities and moving to smaller centres they can both introduce new, labour-saving technology which requires a less-skilled workforce (a process known as 'deskilling') and draw on pools of relatively cheap, non-unionized labour – especially women. Transport costs are of little importance to the production of many goods today, and their increase is more than offset by the other gains. And so industry is shifting to new environments, such as the towns along the M4 corridor in England (e.g. Newbury, Swindon) which have attracted both modern electronics factories and the offices and warehouses of international corporations.

Counter-urbanization is thus a component of the general process of industrial restructuring. Large companies operate three types of operation: their headquarters office, which have to be close to the major centres of financial and political power (although for taxation purposes the companies may be registered elsewhere); their research and development centres, which are usually in pleasant urban environments and in many cases are linked to higher education and research institutions; and their production plants, which are highly automated and employ mainly semi-skilled labour at routine tasks. The last of these are 'footloose', and can be located wherever a compliant, relatively cheap, non-unionized labour force is available. Thus within the United States, the shift of manufacturing to the Sunbelt – especially to the southern States – reflects an exploitation of an available cheap labour pool where unionization levels are very low and where many States operate 'right to work' laws which make compulsory union membership in any workplace illegal; the attractiveness of this area is enhanced by the activities of southern politicians promoting the allocation of Federal funds to projects and contracts in their home States. Internationally, routine manufacturing tasks are being moved from the core of the world-economy (northwestern Europe and North America) to countries where labour costs are much cheaper (as much as 80 per cent cheaper) both nearby (e.g. Spain and Portugal in Europe, Mexico in America) and far away (including Taiwan and South Korea); Eastern European countries are also attractive because of their cheap, disciplined labour forces and the guarantees which the state is prepared to give to outside capitalist interests.

Counter-urbanization is not simply a 'back to the small town' movement, although as people retire earlier, live longer and get better pensions they shift out from the big cities to smaller, pleasanter places, which may also be cheaper to live in, so the size of that movement increases. Fundamentally, counter-urbanization is a consequence of major changes in the organization of industry, from which smaller centres are currently benefiting. There is no necessary relationship between urban size and the availability of cheap, disciplined labour (medium-sized towns were the major centres of unionization and socialism in the USA early in this century). There is one at the present time, however, which accounts for the current patterns.

Problem towns and problem regions

The processes of concentration and centralization have caused relative, if not absolute, decline in the smaller towns of an urban system, as local industries are closed and service establishments lose to competitors from the larger centres. Villages and small towns are the main sufferers and unless they are either close to a big city or can develop new functions – perhaps related to tourism and recreation – little can be done to arrest the decline in their economic and social life and the almost inevitable out-migration of population. Some planning policies have been enacted to concentrate services and employment in 'key' settlements, in the hope of generating some multiplier effects, but in general decay in rural areas and their associated settlements (a decay that continues despite the prosperity of the basic industry – agriculture) has been accepted as inevitable in capitalist economies.

Other casualties of the concentration and centralization processes have been many of the new towns created in the period of industrial capitalism. Most were based on a single industry only, and internalized very little of the growth which was generated by the incomes earned there. They are now casualties either because their resource base is irrelevant (indeed, it may have been worked out) or because the products of their factories – ships and textiles, for example – are either no longer in demand or are being made more cheaply, and perhaps better, elsewhere.

Unlike the decline in the rural areas and small towns, the decline in the industrial new towns is perceived as a major problem in many countries, both because of its extent – the number of people unemployed – and its concentration in relatively few places, which makes it very apparent.

Problem towns and problem regions (containing clusters of these towns) are defined; many of the towns are relatively small, but some (like Liverpool and Glasgow) have grown to large size on a relatively narrow industrial base. The problem, as defined by most observers, is high unemployment, and governments feel constrained to seek policies which will reduce the unemployment, *in situ* if at all possible, in order to sustain their legitimation (the work ethic is strong under capitalism, and without work relative poverty is almost certain) and win electoral support. Thus in many countries of industrial and late capitalism areas have been defined which receive special support for the relief of unemployment problems. The support involves a variety of policies, most of which offer incentives – subsidies for wages, for equipment and for materials; rent-free factories; tax rebates; interest-free loans; transport-cost subsidies – to firms which will establish factories or offices in the defined areas. Similar policies are employed to prevent firms already in the areas from closing establishments and creating further unemployment. Investment in infrastructure, such as new roads, is undertaken to improve the economic conditions affecting firms, and in some countries attempts are made to concentrate new developments in particular locations, known often as growth points or growth poles; the hope is that such concentration will lead to greater internalization of growth through multipliers than would occur if development were more widely spread, and that eventually it will stimulate growth in surrounding towns.

Opinions vary as to the success, and potential future success, of such policies, which like most of the others discussed in this chapter clearly fall into the ameliorative problem-solving category. Certainly there has been some direction of new investment towards the areas carrying the various incentives, probably to the greatest extent in countries which also have strict policies on growth in the big, healthy cities. (To some extent the policies may have been too successful, for some of the biggest unemployment problems are now occurring in the inner areas of the big cities.) But the extent to which growth pole policies are successful in internalizing multiplier processes is far from clear: the evidence from 'natural' growth poles, such as those based on the exploitation of North Sea oil in Scotland, suggests that much of the growth stimulated is not local but in the established industrial and commercial centres. As with all planning policies, however, it is extremely difficult to establish what would have happened if the policies had not existed, and therefore what their effect has been.

Alternative approaches to planning for urban systems

The various urban problems identified in this chapter, and the policies designed to deal with them, are typical of the actions taken in liberal democracies where capitalist influences dominate state activity. Planning in these countries is very largely of the ameliorative problem-solving type; problems are perceived and appreciated, and attempts are made to cope with them without altering the basic forces which bring them about and structure the urban systems in which they occur. Such 'piecemeal' or 'reaction' planning often has a cosmetic impact only, and frequently generates a new round of problems. Occasionally other types of planning are attempted. In Britain, for example, major local governments are required to produce structure plans outlining the desired future pattern in their territories. This often involves only allocative trend-modifying, however, since the existing pattern and ongoing trends are taken as major constraints to future plans; in any case, there is no national plan, only a series of vague outline regional plans, within which these structure plans are set, and the operation of strategic planning by subsidiary local governments ensures that modifications are frequent and are made to accommodate local demands. Similarly, road planning is based on traffic forecasts and involves exploitative opportunity-seeking: attempts to redirect traffic from one mode of transport to another are few.

In capitalist societies, therefore, planning is not strong and probably has had only a marginal influence on the form of the urban system. But what of other countries, where capitalism does not hold sway? Two such societal types were introduced in Chapter 2 – state capitalism and socialism. Each has been built during the last few decades on a mercantile and industrial capitalist base, but the urban framework has been modified within the constraints of the ruling ideology.

State capitalism and the urban system

State capitalism differs from industrial and late capitalism in the ownership of the means of production, but not in its basic goals. Its enterprises are publicly owned and are operated by public bureaucracies, so profit-making is irrelevant. Growth is very relevant, however, and the aim of state capitalism is to achieve high levels of production to create levels of material welfare comparable to those of capitalist countries: materialism is both encouraged yet discouraged – individuals should want to buy, but not to amass.

The main examples of state capitalism are the Soviet Union and its neighbouring countries in Eastern Europe. These inherited an urban system characterized by both central place hierarchies which articulated the mercantile system and major cities towards which much of the surplus was directed. Some of the latter were gateway primates, notably in the Eastern European countries where they were established by Western European or by Turkish colonialism.

State capitalism (which was initially established in the Soviet Union only) introduced industrialization through the development of heavy industries, such as iron and steel, as the basis for economic growth. As in the industrial capitalist countries, new towns were established at resource locations. They were public creations, however, rather than the consequences of private enterprise, and so were established according to a single plan based on a central evaluation of resources: there was a national economic plan with clear spatial components, and labour was directed to employment rather than drifting to it as was the case in capitalist countries. Agriculture had to be developed as well, to support the industrialization programme, but ideological rather than material promises were the main incentives to increase productivity. In several parts of the country, especially in the east, gateway primates were established to 'open up' new farming areas (in many ways reminiscent of the American West), and central place hierarchies were created to provide cultural institutions (newspaper, cinema, library), educational functions (plus boarding establishments where needed), medical services, workshops for tailoring and other essential consumer goods, organizations for agricultural supplies and procurement, and state and party offices.

Soviet cities can be classified according to their economic functions in much the same way as those in capitalist countries, with hierarchies of administrative centres and specialized new towns (such as those containing a variety of manufacturing functions, mining towns, research centre towns, transport centres, resorts, naval bases and so on). Because state capitalism involves a command economy, however, the planning of such an urban system has probably led to a greater number of new towns (over 1,000 in the Soviet Union) and a wider dispersal of industries than was the case elsewhere under industrial capitalism. But there was no coherent urban theory during the early Soviet period. Instead, the theoreticians were divided into those who proposed that industrialization should be based on relatively small urban settlements and those who advocated de-urbanization and a complete urban/rural mixture; to the latter, any form of urbanization indicated the operation of capitalist principles.

The proponents of the urban strategy were the most influential on the development of the Soviet settlement pattern, but although they have been able to achieve much dispersal of industrial activity throughout the country, nevertheless Soviet state capitalism is facing many of the problems of concentration and centralization which face late capitalist societies. The main features relate to the size of the bureaucracy and the drive for high levels of industrial productivity (especially in the production of certain commodities, such as those related to armaments). A characteristic of all bureaucracies is that the power of individuals within them is closely related to the size of their departments, so that the leader of each section requires growth in its area of responsibility; such growth brings with it influence and authority, and success at building up one department may lead to promotion to a bigger, more influential one. Thus the Soviet state bureaucracies are large and competitive, and growth is avidly sought. They are highly centralized, too, despite efforts at regional devolution and decentralization. Thus the national capital, Moscow, is growing rapidly. Many state industries are growing too and contributing to the expansion of major cities, because their chiefs argue that expansion is necessary in order to achieve the desired increases in productivity. Thus although the Soviet Union has an urban system which in its approximation to the rank-size rule is comparable to the American, it faces many of the problems of big cities; Moscow now has a green belt and a series of industrial satellites beyond it. Because state capitalism has many of the characteristics of late capitalism it also has many of the same urban phenomena.

Socialist/anarchist urban planning

The urban patterns and problems of industrial capitalist societies have very largely been reproduced in state capitalist economies, therefore, but in some countries a more clearly socialist orientation has led to different attitudes towards the urban system. Thus in China, in Cuba and in Tanzania policies have been enacted which aim to remove the rural/urban dichotomy within society and to create instead a more balanced situation in which neither dominates.

The Chinese urban inheritance at the time of the 1949 revolution was very similar to that of many other densely peopled, colonially exploited countries: a network of central places articulating local and long-distance trade plus a number of primate gateways (such as Peking and Shanghai) which connected that network to the capitalist world. The urban theory initiated by the socialist revolution aimed to reduce the dominance of the gateway cities and to prevent those industrial new towns necessary to the industrialization programme from growing too large. The emphasis has been on small-scale, labour-intensive production units in both agriculture and industry (except in those science-based industries where advanced technology requires size and specialization); in the service industries, too, specialization has been discouraged so that in the medical field, for example, the emphasis has been on paramedical workers with minimal but sufficient training for the treatment of most minor, frequently occurring ailments and for the encouragement of prevention rather than cure. Overall, the focus is on local self-sufficiency – a form of anarchic communism – although clearly some goods must be provided from specialist towns and some services are far from ubiquitous. The divorce between town and country has been reduced as far as possible to avoid the dominance of an urban administrative cadre, and for several years all urban residents were required to undertake some rural work annually. But the extent to which material improvements are possible with such societal structures, above a certain level, is dubious, and changes introduced during the late 1970s suggested Chinese recognition that a certain amount of individual aspiration to amass wealth was necessary if economic growth was to be extended.

Tanzania also inherited an urban system focused on the gateway primate, in its case the single city of Dar es Salaam; there was a poorly developed central place hierarchy inland from this port centre. As in China, the main aim of the country's development programme (parts of which have been sponsored by the Chinese) has been a reduction of the city's dominance over economy and society, which should create less inequality between people and places in levels of material existence. Thus the focus is on local services attuned to general needs rather than highly specialized and centralized ones serving a small urban élite. Agriculture is to be developed in a labour-intensive mode, with the emphasis on maximizing production rather than productivity. (The latter may under-use land resources in order to return the maximum profit relative to farmers' investments.) Thus the Tanzanian *ujaama* system is anti-urban and aims at rural, local self-sufficiency. The level at which it has begun in most rural areas is very low, and substantial gains in the quality of life are possible before any conflict is created between the anti-urban bias and the need for large-scale production units which will stimulate urbanization.

Conclusions

Capitalism is a dynamic process and so its attendant spatial consequence, urbanization, is likewise undergoing continual change. Such change involves winners and losers; firms may decline in the face of competition, because of the operation of a set of rules based on the survival of the fittest, and so the people dependent on them, in the places where they are located, suffer too, through a series of negative multiplier processes. Other places, with growing industries, prosper, but prosperity and growth often bring with them another set of problems associated with urban size and expansion.

There are bound to be urban problems in a capitalist society, therefore. If the economy operates untrammelled, then adjustments should occur and the urban system will be altered to accommodate the new equilibrium. Such accommodation may come about only slowly, however, if at all. In late capitalism the tendency towards monopoly in many sectors of the economy often prevents the operation of processes which in other circumstances might lead to a new equilibrium, and so the losers remain losers, with their particular problems, and the winners go on winning, with the attendant difficulties of success. This polarization is reflected in the map of the society and in variations within its urban system. To counter it, while at the same time not destroying the basic fabric of the capitalist dynamic, liberal-democratic governments guard the fortunes of the losers while nurturing the activities of any who look likely to succeed. Thus urban problems are tackled by governments. Their usual response, as shown here, is to wait until a problem is widely perceived and the need for action becomes vital (as much for governmental legitimation as for any other reason) and then introduce a solution, one which merely diverts the forces creating the problem rather than tackling its root causes, which may involve tampering with the capitalist dynamic itself. Occasionally, an ongoing trend may be perceived as likely to create a problem and attempts will be made to re-route it. There is no general plan, however; there is no mapped-out future being worked towards, only certain perceived immediate futures to be avoided and present cancers to be removed.

In some parts of the world industrial and late capitalism have not evolved, and instead other socio-economic systems have been grafted onto the existing urban framework. To some extent this has involved the creation of design specifications for future urban patterns which fit the ideology of the socio-economic ethic. In some cases these designs are avowedly anti-urban, and the role of the town and city is kept to a minimum. Such designs produce major constraints on plans for improvements in material conditions for inhabitants, however, and are often not in the best interests of the powerful groups within the society. Thus under state capitalism, despite the general absence of private property and the profit motive, many of the urban problems of the capitalist world can be found. National planning can obviate some of the difficulties of unconstrained capitalism, such as unemployment and underemployment, with the associated drift to large cities in search of any type of work, but some urban problems seem to reappear whatever the local ideology, and appear to be tackled in the same piecemeal fashion.

6. Inside the City: The Rationale behind Residential Segregation

The next four chapters move from an examination of the distribution, size and functions of urban settlements to a consideration of their internal patterning, the mosaic of land uses which is characteristic of all urban places. As indicated in Chapter 1, the nature of this mosaic is a major concern of urban geographers, not surprisingly since the pattern of land uses within towns and cities forms the environment in which a large proportion of most countries' residents live out their daily existence. Although this and the next three chapters cover all aspects of urban land use, most attention is given to the residential areas, which are the primary foci of life, especially social life, for most urban dwellers. This chapter looks at the rationale of socio-economic segregation which is the main feature of those residential areas, showing that this rationale is a reflection of social processes; the following chapter investigates the operation of housing markets as mechanisms for allocating people to different areas of the contemporary city.

Classes and industrial and late capitalism

According to the analysis to be presented here, the major variable determining the residential patterning of cities is the class system. As suggested in Chapter 2, discussion of this system is made difficult by manifold problems of definition. Three possible definitions of classes are available. The first refers to capitalist societies only, and recognizes just two classes – the capitalists and the working class. Such a definition is too broad for present purposes, however, in part because of the blurring of the boundary line dividing the two – by the state, for example, which to some extent makes capitalists of all its citizens, and also by the capitalist class itself, which co-opts many members of the working class into petty capitalist positions in order to win their support for the system's continuation. Thus the division of society is much more fine-grained than a simple capitalist/working-class dichotomy, and residential patterns reflect this graining.

At a second level, classes are sometimes associated with individual economic categories within the division of labour, such as doctors and lawyers, motor mechanics and train drivers. Each of these groups has its own vested interests and identity, both in many cases represented by a trade union, but such a detailed categorization is unnecessary for the present purposes, as also is one which divides households and individuals into a large number of groups on the basis of income. Thus the focus here is at a third, intermediate, level, at which society is divided into a number of major categories. The boundary lines of such categories are almost impossible to clarify, but the following general classification is sufficient for the analyses here:

(1) the major capitalists – the owners of large industries and service establishments with substantial workforces;
(2) the minor capitalists – employers of small numbers of workers in shops, offices and other establishments, plus the self-employed;
(3) the major professionals;
(4) the managers of various enterprises, including state bureaucracies;
(5) the minor professionals;
(6) the minor white-collar workers, such as secretaries and employees in shops;
(7) skilled artisans;
(8) semi-skilled blue-collar workers, many of them factory operatives; and
(9) the unskilled.

Some of these classes of workers are more numerous in certain countries than they are in others. State capitalist societies, for example, have no members of the first, and the second may be accepted but not encouraged; the fourth class, on the other hand, will be very numerous.

An important function of the class system is to act as a procedure for allocating the rewards of a capitalist society. There are three types of reward. The first is income, and its associated accumulation – wealth. Income is a payment for employment which varies between occupations and classes; in broad terms, the lower the position of an occupation on the scale just outlined, the lower the average income is likely to be. (There are two major exceptions to this. The first refers to the skilled workers in class 7, many of whom earn more than those in class 6, especially the many women who work in offices and in shops. In many societies, however, such minor white-collar occupations are accorded greater status than are skilled artisan jobs, even if they are less well paid, and a white-collar occupation is an aspiration that many blue-collar parents have for their children. The second refers to the fact that capitalists do not receive income in return for their labour power; they receive profits from investing in the labour power of others.)

Income differences reflect differences in the price paid for labour, in exactly the same way as the prices of other commodities vary in a capitalist system. Some occupations may receive relatively high incomes because their workers are more productive than those in other occupations; in some, high incomes reflect a high general valuation of their work by society as a whole. But the major determinant of wage levels, as with all other prices, is the ratio between supply and demand: occupations for which labour is in relatively short supply but is in great demand are likely to be among the most highly paid. A major route to a high income, therefore, is entry to an occupation which is both highly regarded by society and has a favourable supply: demand ratio (from the worker's viewpoint).

The second reward is power, the ability of an individual or group (class in many cases) to manipulate society towards certain ends. To some, power is itself a desirable end, for they gain considerable satisfaction from its exercise. But for most it is the means to the end of greater incomes and wealth.

Finally, the third reward is status, a largely intangible indication of the standing of an individual, a family, a group or a class within their society. (Measuring instruments are available which provide at least an ordinal ranking for status.) Again, this reward may be sought as an end in itself – certainly it is widely sought. But most people require status not just for the charisma that it brings but also for its relationship with the other rewards. Thus, for example, people with high status are often deferred to by others, and their views are given disproportional weight in decision-taking; this in turn gives them power, which can be employed to obtain income and wealth. (Except of course in state capitalist and socialist societies, where income differentials are supposed to be much narrower than they are in capitalist societies and where the accumulation of wealth is in general prohibited. In these, power and status may have to be accepted for themselves.)

These three types of reward are interdependent, and possession of one can be used to advance claims for the others. Why, then, are they distributed unevenly between the various classes, with income, wealth, status and power generally greatest in the upper strata? (There are exceptions to the general correlation just outlined: in many capitalist societies, for example, occupations associated with the religious functions carry much status, not a little power, but only small incomes; often the status is retained after both the power and the relatively high real income once enjoyed have been removed by secularization processes.) The answer lies in initial advantage. Initially, wealth, status and power were vested in the landowning class – through their own efforts. As mercantile and then industrial capitalism evolved, so some of the wealth was redistributed (by the capture of the new wealth by the new classes rather than by any major removal of wealth from the landowners), and the power that accompanied it was used to ensure an allocation of income that benefited the expanded capitalist class and their offspring. The occupations of the latter – such as many of the professions and the managerial tasks – were granted both status and substantial incomes, which brought power and an ability to manipulate the future class system. As the division of labour proceeded, so those in power were able to control the definition of status and the allocation of income to new occupations, with only occasional challenges to their creation of the society's form. Simultaneously, supply: demand ratios had to be manipulated, particularly in the occupations of the most powerful, in order to ensure that no imbalances would arise to affect incomes and challenge the distribution of power. It is this manipulation, especially of supply, that characterizes the year-to-year operation of the class system, and accounts for much of the rationale behind the residential patterning of towns and cities.

The operation of the class system

The classes identified above are employed in the analysis here because of the interests common to each, which both give them internal cohesion and set them in conflict and competition with each other. (There is also competition within classes for rewards – between doctors and lawyers, for example – but this is of marginal relevance to the study of urban residential patterns.) These common interests are reflected in a variety of ways, such as the political parties to whom the members of the various classes allocate their votes.

The basic common interest within each class involves ensuring the ability of each individual to maintain his position within the class system, to ensure that his heirs enjoy at least the same position, and, if possible, to improve that position by movement upwards through the class system. For those who consider that they are in the highest class, the last aspect of the basic interest is irrelevant, but they are probably concerned to improve their relative position by obtaining a larger proportion of the society's income, wealth, power and status for themselves. For all other classes, except those at the bottom of the system, this 'self-protection' involves looking 'both ways at once', seeking betterment through movement upwards within the class system while at the same time resisting erosion of their relative status, income levels and power by those seeking to move up from below them.

Maintenance of class position is easiest if mobility is impossible and the society operates via prescription, by the allocation of the various positions within society to particular families and their descendants; this is furthered if inter-class mobility is also prohibited, with no inter-marriage, for example. This is the situation in rank-redistribution societies with rigid class boundaries that are virtually impervious (totally so in certain caste-based societies). Such a situation benefits the most powerful class to the greatest extent, of course, since their position is unchallengeable, except by total revolt: others are protected from 'invasion from below' but also prohibited from aspiring to upward movement.

Such a rigid class system can operate only in a conservative, static society, and even the creation of new classes (castes) is unlikely to allow rapid social and economic change. This rigidity is inimical to the goals of a capitalist society which emphasizes competition and rewards for enterprise and which stimulates these through the promise of gains to be made from work. A capitalist society must allow inter-class movement, by dangling the potential rewards of income, status and power before those who would seek self-improvement, thereby both increasing productivity and consuming the increased volume of goods and services produced. At the same time the competition stimulated within a capitalist society is unequal, so that the occupants of each class can repel, as far as is both necessary and desirable, the incursions of others. Such repulsion requires class cohesion and activity because individuals are unable to manipulate the society themselves without the support of like-minded others.

There are three main routes to the sources of income, status and power in capitalist societies, and thus to inter-class mobility: they are wealth, position and skill, and their relative importance varies between classes. Wealth allows individuals to become capitalists and to gain control of a proportion of the means of production, which in turn creates greater wealth for them. Most wealth is inherited and accumulates as it is re-invested by each succeeding generation. Much of it is held either as land or as other forms of fixed capital, such as machinery and buildings.

Position implies that some people have greater access to certain job opportunities than do others, because of who they are. Nepotism is an obvious example of this, as is the currently more important practice of patronage; many positions, in all types of society, are in the gift of an individual holding a particular post. And, finally, there are what are generally known as the 'old boy networks' – institutionalized in such activities as freemasonry – through which people obtain positions via their personal contacts, such as those who attended the same school or university as themselves (or their parents).

Wealth and position are undoubtedly important access routes to a variety of occupations, especially at the upper end of the class system but not exclusively so with respect to position. But in a capitalist society, especially in its late capitalist stage, built on technology and on investment in machines for the increase in productivity necessary to growth, it is skill (or qualification) that provides the major criterion for entry to, and continued residence in, most occupations. This investment in technology applies not only to manufacturing and agriculture but also to many parts of the tertiary and quaternary sectors: where machines are not used in the latter, other skills – of analysis, interpretation, communication and manipulation, for example – are needed. Few of these are innate; many are highly specialized and must be learned, in some cases only by long years of training and practice. The education system which provides that training is thus the key to entry to an occupation and the foundation of a late capitalist society; much inter-class conflict has been waged over access to educational opportunities.

Both wealth and position have been important means of obtaining access to educational opportunities, and thence to the more desirable occupations. They still are, especially wealth, but in less direct ways than was the case at earlier stages of capitalist evolution. At one time members of the higher classes maintained their positions by providing educational resources for their own children alone. These resources had to be bought and so were not available to the large mass of society with insufficient incomes; they allowed a small amount of social and economic mobility within the upper classes only. Increasingly, however, the demands of industrial capitalism required a basic level of education for the members of most classes. Initially this was provided in relatively cheap schools whose fees could just be met by the customers, who were able to invest in their children's futures and to guarantee for them relative success, compared to those from the lower classes whose parents were unable to invest in their education. The boundaries of the class system were maintained by differential provision of educational opportunities and of access to them, whilst the requirements of an expanding capitalist mode of production were met by a movement of the whole class system upwards on a general scale of educational achievement.

As the need for more education was appreciated, and as it was demanded by the lower classes who realized the entrée which it gave to better-rewarded occupations, so its provision was taken over by the state as part of its collectivization of the means of reproducing labour power and externalizing the costs of capitalist enterprise. Universal education was achieved, and the length of the period of compulsory education slowly extended. But this has not provided equal opportunity for all to enter every occupation, for a variety of reasons. To protect the higher-status, better-paid occupations their skill and qualification requirements were steadily increased, demanding education either beyond the level provided free by the state or beyond the age at which families could afford not to have their offspring working and contributing to the family income. Thus the members of the upper classes retained the greater access to professional and managerial occupations that their children had always enjoyed. Some did this by the purchase of education outside the state system, their expenditure not only gaining access to superior educational resources and greater probabilities of obtaining entry to more prestigious universities, for example, but also ensuring positional benefits, through the operation of old boy networks. For those in the lower classes, however, it was necessary to manipulate the state system, in ways which will be described below.

The expansion of industrial and late capitalism has produced an increased relative demand for labour in the various classes which require particular skills especially in those involving professional, managerial and certain artisan occupations. While striving to protect their own positions, and those of their children, therefore, the members of those classes have had to open their boundaries somewhat, allowing upward economic and social mobility but carefully controlling the rate of entry, so as not to imbalance supply: demand ratios. Education has been widely perceived as the vehicle providing access to social mobility – especially for those lacking initial wealth or position – which accounts for the conflict between classes over access to educational and other resources which is strongly reflected in intra-urban residential patterns.

Immigration and urbanization

The ethos of industrial and late capitalism includes a strong commitment to expansion, and although improvements in productivity release many of the demands made for certain skills and allow movement upwards of certain of the lower classes, it is necessary to maintain a large membership of the lower classes, if only to ensure that the supply: demand ratios do not shift too strongly in the workers' direction and fuel their demands for higher wages. This requirement for growth in the classes characterized by small amounts of income, status and power, whilst at the same time encouraging aspirations to, and some achievement of, upward economic and social mobility (if achievement was not visible, the aspirants would soon be frustrated and led to dissent), is met by urbanization. Some part of this is a product of population growth *in situ*, but migration plays a major role in capitalist urban expansion.

Migrants to towns and cities come from two main sources. The first comprises those areas whose prime functions within the spatial division of labour are characterized by a declining demand for labour consequent upon its replacement by machines. The rural areas, dominated by agricultural pursuits, have thus been the major source of migrants to the expanding cities, followed more recently by the smaller towns, where the few industries are shedding labour and there are not alternative sources of employment available. The second major source has been of migrants from similar types of origin, but in other countries, reflecting the international division of labour under industrial capitalism. For these migrants the opportunities offered in the cities of more advanced industrial countries are perceived as so much greater than those at home that they emigrate – many of them intending the move to be only temporary, as a means of amassing wealth, but eventually effecting a permanent change of residence and even nationality. The rapidly expanding industrial nations with few local labour reserves, such as the United States, have been the main destinations for such migrants.

Some of these migrants, of both types, have particular skills to sell and move to partake of the perceived opportunities at their destinations. Most are both unskilled and lacking in wealth, however; they move for the promise of social and economic improvement in an alien environment – except for those forced to migrate, by religious or political persecution, perhaps, or to escape famine. Many have no prior arrangement made that will allow them to partake of the wealth on their arrival, but migrate in the hope of material gain for themselves and, in many cases even more so, for their children. In an expanding economy a large number will probably succeed in finding an economic niche that provides more security than they had achieved elsewhere, even though the rate of social and economic advancement is neither as great nor as rapid as they had hoped. But in a relatively stagnant economy (as in much of the industrialized world at present) they may be frequently, if not always, unemployed, because of the lack of demand for their few skills. Their economic and social advancement may then be very slight (although social benefits, such as health and educational services, may be available for their children); this is the chronic situation facing the migrants to most of the cities of the Third World, where capitalist expansion is insufficient in format (because it is capital- rather than labour-intensive) to provide jobs for all.

The arrival of migrants to occupy the lower strata of society allows the existing residents to protect their own positions. The newcomers, especially those from alien backgrounds and cultures, require time to accommodate to their changed situation, during which they pose little threat to their hosts (especially as the migrants will be prepared to work within the system rather than to challenge it and risk repatriation). As accommodation proceeds, so the migrants begin to seek advancement. An expanding economy will have gaps for them to fill, while the positions that they vacate are occupied by newer generations of immigrants. Thus the whole social system shifts upwards, and the migrants remain excluded from entry to the higher levels, to which they almost certainly aspire.

Upward mobility for members of migrant groups, by which they move into the strata formerly occupied exclusively by their hosts but now to be shared with them, is frequently slow. Assimilation of the immigrant groups, implying their loss of all individual and group characteristics and total absorption into all aspects of the host society, may take several generations, if it is achieved at all. Mechanisms to protect the relatively privileged economic and social positions of the hosts will be devised, and the immigrants confined to strata where they occupy the menial positions that established members of the society shun. Such barriers to entry into various occupations are most readily operated if those to be discriminated against are easily recognizable; skin-colour is an obvious basis for discrimination, therefore, which leads to the fostering of interracial prejudices and caste-like status for certain immigrant groups. Such migrants will be partly integrated with the urban society, because their labour is in demand, and their allegiance to the system may be sought through the granting of certain economic benefits. They may be denied assimilation, however, in order to prevent them from competing for the higher-status occupations and to maintain them in a subservient social and economic position: indeed, many countries now impose strict immigration quotas to protect their own residents from labour-market competition, and several prefer to import temporary labour only (involving workers who must leave their families 'at home'), which can be repatriated when demand slackens and the immigrants threaten to become a drain on local social services.

Summary

Two major themes have been developed in this introductory section. The first relates to the role of education in both industrial capitalist and state capitalist societies. Economic growth is dependent on technological sophistication to propel the capitalist dynamic, which is based on increasing the productivity of labour. To organize and operate this technology, the population must be divided into groups, each of which is provided with sufficient education to undertake the requisite tasks asked of it: each educational level is associated with occupations that carry with them certain incomes, status levels and power. Education is the passport to success, therefore; all want this passport and so compete for educational opportunities. (Some individuals, and even groups, may be partially alienated from their society and its dominant goals, and so do not desire education and the potential for social and economic advance that it offers. This is of benefit to others in society since it reduces the competition for some jobs and helps to maintain a supply:demand ratio which keeps incomes low in the lower strata of the class system. Partial alienation may even be encouraged by members of the higher strata.) Some must lose in the competition for education and jobs, at least relatively; those with initial power will attempt to ensure that they and their children have better access to the opportunities, and thence to the rewards, than do others, thus maintaining their relative positions within the class system.

The second theme concerns the role of immigration in urbanization and the expansion of capitalist societies. Immigrants move to the cities to find jobs, most of them in the lower strata of the class system where vacancies occur because of the social and economic mobility of the established residents (the hosts). Most immigrants aspire to participate in that mobility. Some are successful, especially those able to take advantage of a period of economic expansion when opportunities are plentiful. Of the many immigrants who may be present in a large city (including those from other countries and cultures), the most successful are usually those who differ but slightly from their hosts, for whom assimilation of these newcomers involves few traumas. Others are not as fortunate, and their differences may be used by their hosts as the basis for discriminatory practices, many of them bred out of fear of what inter-cultural contact may promote. Such discrimination usually condemns the immigrants to permanent occupancy of the lower strata, and because those strata suffer most in periods of economic recession the immigrants often bear much of the brunt of periods of stress. Such stress may have a variety of representations: xenophobia is common during recessions, for example, and may be stimulated by political actors seeking scapegoats foı social problems.

These two themes, of conflict between classes over educational opportunities and between immigrants and their hosts, are crucial to the discussion of residential patterns which follows in this chapter and the next. They are not the only factors that influence the form of the residential mosaic, but income, wealth, status and power – all of which are linked to education and to migrant status – are major determinants of the behaviour patterns that lead to the spatial segregation of classes.

Distancing and residential location

Some of the inter-class competition and conflict discussed above has been controlled by the upper classes, through *de jure* and *de facto* discrimination in the labour market. Certain occupations may be reserved for people from particular backgrounds, for example, as in India's caste system, the concept of reserved occupations ın South Africa's apartheid and, less rigidly, the Oxbridge dominance of certain grades of the British Civil Service. Similar barriers to immigrant assimilation can be imposed, too, as was the case in the United States with black slaves and in many countries with regard to Jewish settlers. Such legalized discrimination reflected the close-to-absolute power of the upper classes, but as this was slowly yielded under pressure and economic and social mobility was allowed (the Civil Rights campaign in the United States illustrates such yielding) so alternative mechanisms were sought which would allow classes to protect their interests from invasion by those of lower status and background.

One of these alternative mechanisms is distancing. In a relatively static society, with assured social and economic status and very little social and economic mobility to challenge the established order, members of each group within urban society 'knew their place' and acted accordingly. Apprentices and craftsmen lived in their employers' homes, frequently as members of the household; servants, too, lived on the same premises as their masters and mistresses, albeit on separate floors of the dwelling and almost always in much inferior conditions. Their relative proximity to the higher-status groups was no challenge to power and status, since they had no basis to hope for entry to the higher echelons of society.

The early town, prior to the onset of industrial capitalism, was not characterized by marked residential segregation of the classes, therefore. Indeed, in the early development of industrial capitalism many employers built estates of homes for their workers, alongside their own and those of the managers: the differences between the classes were reflected in the size and quality of their dwellings, but not their addresses. It was only with rapid industrialization and urbanization, reflecting the greater mobility potential, that the upper classes began to move away from their employees, in part to protect their economic and social positions (other reasons included protection from perceived health hazards of living among the poor). Thus began the process of social distancing, which not only led to separation of the upper and lower classes but also to a great deal of fine-grained sifting of residential areas, reflecting the increasingly complex economic and social division of labour.

Distancing and externalities

The earlier discussion of industrial growth in cities introduced the concept of external economies to account in part for factory agglomeration. External economies involve firms benefiting from aspects of the local environment – such as other firms producing the same product; neighbours who both provide needed inputs and consume outputs; an infrastructure of services and facilities; and so on. Because of these economies, firms are more efficient than they would be if they operated from isolated locations, and this is reflected in their profitability.

The concept of external economies can be extended to cover a variety of economic and social relationships, and is important to an understanding of the rationale for residential separation of classes. In this context, external economies are generally known as externalities, which are defined as aspects of the local environment which contribute to the quality of life of an individual, family or household resident there, but which are not purchased directly by them. Such contributions may be considered either good or bad, so there are both positive and negative externalities. A beach or a park in the neighbourhood may be considered a positive externality by most people, therefore, as might a view of distant mountains: each adds to the pleasure of living there. An industry, particularly one which polluted its environment, would be widely considered a negative externality, on the other hand, since it would detract from the pleasures of living nearby.

There is a broad consensus over some aspects of local environments, so that most people will define them as either positive or negative externalities accordingly. Polluting industries fall into the latter category, and so areas close to them are generally considered as undesirable residential locales, to be avoided if at all possible. In capitalist societies many homes are purchased, and so their desirability is reflected in their price. Thus the fewer the negative externalities in an area, the more desirable its homes and the higher the price that those with the available money will be prepared to pay for them. As a consequence of this, the lower one's income the more likely it will be that the only homes that can be afforded are in areas with considerable negative externalities; dwellings close to polluting factories, especially those downwind of them, will be 'reserved' for the lower classes since these are the only properties they can afford and no other classes wish to live there. Exactly the opposite effect occurs with positive externalities. A south-facing hillslope (in a northern hemisphere city) is likely to be more attractive than the valley floor below it, which houses the city centre and which is probably damper and more prone to both frosts and fogs. As a consequence, those with most money available for purchasing homes will wish to live on the slope, and will bid up prices here beyond the capabilities of their poorer counterparts in lower classes. Thus greater incomes allow members of the higher classes to reserve the pleasanter environments for themselves and to confine their lower-class contemporaries to areas with many negative externalities, such as air pollution, noise, traffic congestion and health hazards.

The physical environment provides examples of positive and negative externalities which are widely perceived, not only in industrial capitalist societies (as in John Braine's *Room at the Top*) where properties are bought and sold, but also in state capitalist societies where other mechanisms are employed to allocate people to housing. But it is externalities in the social environment, with their implications for the maintenance and improvement of one's societal position, that are generally of more relevance for the understanding of residential location patterns. Such externalities were recognized by Michael Frayn (*Sweet Dreams*, Penguin edition, 1973, p. 40):

> The children are at school. The school is many-windowed, relaxed, cheerful, with a good social mix and high academic standards. It's reached by a pedestrian walkway through the treetops, well away from all roads. They get their children into it by a miracle.

The externality to which Frayn alludes involves the desire to ensure that the lifestyles, public behaviours and social and economic habits of one's neighbours are at least consonant with one's own; they should certainly not be worse (and therefore characteristic of a lower class), and if at all possible they should be better (characteristic of a higher class).

For many people this neighbourhood consonance with their own values is especially important for the raising of their children. As already argued, most parents wish their children to obtain a place in the class system which is at least equal to their own; the children must be socialized into the correct habits, and must receive the requisite education that will both instil those habits and provide the requisite skills for economic and social success. A neighbourhood consonant with these goals is widely believed to provide an environment conducive to their achievement.

Neighbourhood and socialization

A child's socialization involves its observation, acceptance and emulation of the models available to it. Initially, those models are provided by its immediate family, especially its parents, and most of its early learning and behaviour is home-based. Later, most children come into contact with others living locally, either formally (perhaps through a parental introduction) or informally (by chance meeting in a public place). Since most parents believe that social and economic success in a capitalist society is dependent, at least in part, on observing and emulating correct patterns of behaviour, they will wish their neighbours to be of a like mind and to conform with their own definitions of correctness. Contact between neighbourhood children will then be mutually reinforcing of the norms instilled in the home environment.

Outside the home, the major impression on a child's socialization is made by the school. There exposure is to a much larger number of models, which potentially may range widely from those reinforcing the norms of the home to those which run counter to them. The parent who hopes for economic and social success by a child fears that contact with the wrong models, especially prolonged and widespread contact, may lead him or her in directions other than that wished. Clearly, these undesirable contacts are likely to be with children from lower economic and social classes, who may have different outlooks on life, not to mention their possibly 'unacceptable' behaviour; the children of parents from higher classes, on the other hand, will introduce desirable new models which can only have a positive effect on the child. A school which provides the undesirable models in any volume is thus a negative externality, whereas one in which the behaviour patterns conform to the parents' own is a positive externality.

Almost all educational systems are locally based, so that children attend the school nearest to their home; for primary schooling (up to the age of about eleven) this school may be close to the home and act as the centre of a variety of local organizations. Thus living in an area whose residents have social and economic backgrounds similar to one's own should result in one's children meeting others from similar homes in classroom and playground. As a consequence, parents who see the school as a major avenue of child socialization will be prepared to pay extra costs to live in an area whose neighbourhood school will reinforce their educational requirements. If most parents have the same outlook, then the result will be segregation of households, and thus classes, on the basis of income, undertaken to purchase schooling of a defined quality.

The school introduces a variety of habits to the child, such as attitudes to learning, to the role and importance of education and to career aspirations. Schools which provide higher-class social environments are thus likely to introduce attitudes to education which are oriented towards achievement and success. Most teachers come from relatively high-class backgrounds, and more easily establish a rapport with children from similar origins and with the same attitudes and desires with regard to education. They prefer to work in schools in higher-class neighbourhoods, which not only are academically oriented, thereby both offering a certain type of job satisfaction and providing a basis for career advancement, but also pose fewer discipline problems. In a relatively free market, in which teachers apply for vacancies and schools choose those they consider to be the best candidates, it is the schools in the higher-class neighbourhoods which are likely to attract the better teachers and provide the most stimulating academic environments for their pupils. In lower-class areas, where the school may have to overcome a relatively apathetic home environment if a child's potential is to be fulfilled, because many parents prefer the immediate gains of early employment to the promise of returns after a longer period of further education, the relatively poor resources may fail to realize the educationalists' aspirations. (This may be of value to the higher classes, since a lack of aspiration and achievement-orientation by the lower classes protects the former from any major competition from the latter.)

Distance and interaction

For parents themselves, and for others without children, a neighbourhood consonant with their own background, tastes and attitudes also provides a positive externality-whereas one which contains households of a higher status is often considered a bonus. Social benefits stem from this externality. People make acquaintances and develop friendships at a variety of times and places during their lives. For many, their neighbourhood may be the source of many of those contacts, and since most lasting friendships involve people who are social equals and have similar backgrounds and interests a homogeneous population of like-minded people is attractive as one in which to develop roots. An area in which many of the residents are unlikely to offer valuable social contacts is likely to be considered undesirable, on the other hand, and even classed as a negative externality if the public behaviour of those residents (noise, litter, unkempt gardens, etc.) offends.

It has been widely canvassed from the results of social scientific investigations that distance is a major constraint on social interaction. The evidence for this comes from surveys in which people are asked where their main friends and acquaintances live. Most respond that they live nearby, and graphs show very clearly that the greater the distance from a person's home, the smaller the number of friends and acquaintances resident there, despite the increased number of opportunities for interaction. (This finding applies over even very short distances, such as within a cul-de-sac or along a single floor in a block of apartments.) Such findings lead to determinist interpretations to the effect that because of the costs (in time as well as money) of moving about, most people are constrained to make social contacts in a restricted area around their homes. But it was argued earlier that people select an area because its residents have attitudes consonant with their own and so are the sorts of people they want to make contact with. The pattern observed by researchers may indicate a conscious structuring of distance by residents, therefore, and not an effect of distance on individual behaviour. Nevertheless, although most people develop acquaintances on the basis of common interests, location is a common constraint, especially for those who are relatively immobile – such as the old, the infirm, the poor, mothers with young children, and those lacking access to various forms of transport. For these in particular, the neighbourhood provides the major, if not the only, set of opportunities for social contact. For others, the space that they can search for lasting friends may be wider, but many start from their home and seek outwards (others, of course, start from an alternative base, such as the workplace); some may have to search long and far before discovering satisfactory, sustainable relationships, but the research findings indicate that most are able to make viable contacts with their neighbours.

Not everybody enters an area in the expectation of making social contacts there, and indeed quite a lot of people have no local friends and acquaintances: this is particularly the case for adults without any children in their household, for children are a major catalyst of inter-parental contact. But even the social isolates (in a neighbourhood sense) usually consider it desirable to live in an area whose residents' characteristics are consonant with their own. This is because such people have the same norms with regard to public behaviour and neighbourhood character, and will be prepared to act to maintain the area and protect its character and positive externalities. The neighbours may not be in communion, therefore, but they form a community of interest, albeit probably an unrecognized one whose potential is only realized in the face of a threat.

One further aspect of neighbourhood social interaction and externalities concerns the continuation of the class system. This is in part guaranteed, as far as possible, by manipulation of the educational system. It may be furthered by marriage. If both bride and groom are from the same social class, then the next generation is likely to occupy that same class. Thus many parents are concerned that their children do not marry 'beneath them'; marrying 'up the social ladder' is desirable, but can refer to only one half of the marriage, and so most marriages are intra-class. In earlier societies parents were able to manipulate the future generation by arranging their children's marriages. This is now rare in capitalist societies. To try and ensure that 'desirable' marriages are made, therefore, parents manipulate social relations to create situations in which their children come into contact with acceptable potential spouses. Consonant neighbourhoods offer an environment for this. Many marriages are stimulated outside the home area, of course (on holiday, at universities, through work, etc.), but detailed evidence suggests that a large number of brides and grooms are near-neighbours before their wedding. Whether it was residential proximity which brought them together cannot easily be established, any more than it can for friendship creation, but circumstantial evidence suggests that many parents' desires concerning their offsprings' marriages are satisfied through their choice of residential location.

Perhaps whether the neighbourhood has an important influence on various types of social and economic behaviour is not particularly important in understanding the genesis of urban residential segregation. What is crucial is that many people believe that it does, and so they define positive and negative externalities accordingly, and attempt to find a residential location on those bases. Although some positive externalities are the same for all, with others what is a positive externality for one is a negative externality for someone else. And since people pay to live in areas from which negative externalities are excluded, the relatively poor are in many cases denied the opportunity to obtain their defined positive externalities. The result is residential segregation of the classes: the lower the incomes and power of a class, the more likely that they live among their peers only and are unable to purchase any positive, social externalities.

Property values

The price of property in an area reflects evaluations of its social environment and the balance of positive and negative externalities there; people are prepared to pay high prices to avoid undesirable areas. Because certain externalities refer to whole neighbourhoods rather than to individual properties (a primary-school catchment area may be coincident with a neighbourhood's boundaries, for example), this produces a clear differentiation between areas in the values of properties, even of equivalent properties in terms of age, size, type of materials and so on. This map of the consequences of perceived externalities is available to all, in part (through property advertisements and public records of sales) if not in total.

In a capitalist society which is based on private ownership of goods, including land and housing, the price paid for a home reflects not only its value as a shelter and as a place in which to live but also its investment value. The positive externalities (or, alternatively, the relative absence of negative externalities) in an area influence the size of the investment. However, large it is, the investor will not wish to see the size of his equity capital reduced and will aim to see it increase, at least at a rate equal to the general level of inflation in the society. He will thus fight, with his neighbours, to maintain the existing positive externalities and will seek to extend them; more importantly, he will resist, as strongly as possible, any attempts to introduce negative externalities which might cause his property's value to decline, relatively if not in absolute terms.

Once a pattern of residential segregation of the various class groups has been established, therefore, vested interests will oppose all proposed alterations except those which bring positive externalities only. And yet change is almost certain in any growing city. New industries and factories will be established for which sites are required; new transport routes will be needed to maintain the city's efficiency; new housing areas will be developed; and so on. All of these location decisions are likely to bring some negative externalities, at least to some people, in their surrounding areas, and they will be resisted by those whose property values and social environments are threatened. As income generally carries with it a greater ability to purchase access to positive externalities plus a greater potential loss from the impact of new negative externalities, the higher-class groups are likely to exercise their relatively large amounts of political power in the conflict over the location decisions. And they are the probable winners in such uneven competition. As a consequence, new negative externalities are most likely to be located in, or close to, the areas which already have most of them, thereby accentuating the differences between residential areas, which are reflected in property values.

Because changes to the urban fabric introduce new sources of positive and negative externalities, they are potential generators of local conflicts. In the face of such proposed changes, the main protestors are usually those with most to lose: property-owners, who perceive possible falls in land values, and parents, who identify potential deterioration in an area's schools. In general, it is the more affluent property-owners who have most to lose and who, because of their ability to purchase legal and technical advice and their greater knowledge of, and links to, the political systems within which such conflicts are adjudicated, are most likely to prevent changes likely to injure their interests. Such conflicts are usually played out locally, but their existence is part of the dynamic of capitalist cities. Alterations in land use are needed if investors are to achieve profits, and if the losers in the conflicts over changes are the less affluent, then the price paid for those changes is substantially carried by them. Local conflicts are part of the general contest between classes within capitalist society.

Immigrant groups and distancing

Most immigrants to expanding cities occupy the lower strata of the class system. As such, they are priced out of most of the housing opportunities and are constrained to living in the least desirable residential areas which contain most negative externalities. Thus a consequence of their class position is their residential segregation relative to the city's population as a whole.

For most immigrant groups this class-based argument provides only a partial account of their residential segregation. With relatively few exceptions (usually those groups which do not differ markedly from their hosts in racial and cultural features), such groups are in the lowest classes and so occupy the least desirable neighbourhoods. The degree of residential segregation between them and those of their hosts in similar occupations, and also between them and members of other immigrant groups, indicates the operation of further mechanisms which involve distancing of cultural and ethnic groups one from another. Several reasons have been suggested to account for such distancing.

The first relates to prejudice within the host society and its reflection in discriminatory practices aimed at keeping all social contact between themselves and the migrants to a minimum. The major cause for such antagonistic attitudes and behaviour is thought to be fear of the unknown, which receives particular attention at times of economic recession when the migrants are presented as competitors for a small number of employment vacancies and as recipients of welfare benefits paid for by their hosts. Anti-immigrant feeling, directed mainly against visible groups such as those with different skin-colour, is common during such periods. But the fear is more widespread, and many myths are generated concerning the newcomers' behaviour (regarding health arrangements and sexual activity, for example) which make them out to be unsatisfactory neighbours for their 'respectable' hosts. (Throughout this section, the general term 'immigrant' is used for both those who have arrived from another country and those whose origins are elsewhere in the country containing the city to which they move. The latter are much more likely to be of the same cultural and racial background as their urban hosts, and to suffer no more discrimination than others in the same class position; the major group of intra-country migrants suffering from discrimination is the black population of the United States.) Distancing is brought about as a result of the fears which are based on the racist myths, and the lower-class immigrants find themselves constrained to a few residential areas only, which their hosts refuse to share with them.

Immigrant groups are in general labelled as strong negative externalities, therefore, to be shunned by members of their host society. This labelling may stimulate a second cause for their segregation – a desire among the immigrant group to retain a separate identity, maintaining their cultural traditions in an independent social enclave whilst participating in the wider urban economy. Most groups display such cultural cohesiveness most strongly when they first arrive in the new city, as a protective device against the alien environment, many of whose features they do not understand. Cultural cohesiveness not only allows the group to retain its native characteristics – desirable to many members because of their intention to return home rich – but also provides an environment for group learning about their new home. The attitudes of hosts and immigrants may thus be mutually reinforcing.

The host society commits the newcomers to certain neighbourhoods, therefore, in much the same way as it commits other outcasts to prisons and asylums. Those neighbourhoods then develop their own social systems, perhaps with class differences that are reflected in residential patterns within the ghetto to which the migrants are confined. The cultural separateness which this fosters may continue long after the host society is prepared to accept members of the group as their economic and social equals, when conflict and distancing can be replaced by mutual accommodations. Continued discrimination and confinement to ghettos may breed major social discontent among the migrants, however, and lead to outbursts of inter-group conflict along the border of the rival territories. Gradual assimilation, and slow movement of some of the migrants into the society's higher strata, may counter such a tendency. But even with widespread acceptance among the hosts, some of the migrants may choose to retain their separate identity and residential area. They may act as an entry-station for new migrants, operating what is known as a chain migration process. The pioneers send back reports of success in their new homes, often accompanied by remittances of money, and encourage family and friends to join them; the sponsors may well find jobs and homes for the new arrivals, usually associated with their own. Thus some 'urban villages' are established, with social structures similar to those in the homeland; such villages are very common among the Southern European settlers in the cities of the New World of America and Australasia. The cohesiveness of these units maintains the cultural and spatial distancing of the group and provides a staging post through which later arrivals may pass en route to full assimilation with their hosts.

Conflict between different migrant groups within a city can also be generated through fear, ignorance and myth. These groups are competing among themselves for acceptance by their hosts and for economic benefits, and their competition may breed conflict. (Such conflict may be, at least indirectly, encouraged by the host society, to deflect any latent challenge to their own power and status.) The existence of such conflict may generate its own extension, as it provides an extra dimension to the pride which individuals feel in their group identity. Endemic sniping between groups may be the consequence, with each defining its separate territory that has some of the characteristics of a frontier outpost in a hostile land.

These two types of conflict, between immigrants and hosts and between different immigrant groups, are exaggerations of the more general inter-class competition which occurs in capitalist societies. All classes and groups have their interests which they seek to protect. Competition is at its fiercest when it is closely allied with economic status, when it is not merely a contest but a challenge, a threat to one's status and standard of living. Thus mechanisms are required, as in the educational system, to protect those interests. If the threat can be clearly defined, because the challengers are readily identified, then it is relatively easy to achieve consensus among those threatened: the spread of argument and propaganda should ensure recognition that the threat can be muted by residential segregation of the assailants. Thus members of the migrant groups are accepted into the economic structure because their labour power is necessary in its lower levels. But they are denied the access to those positive externalities which might allow them to infiltrate the higher strata. (They can, of course, create their own higher strata within the ghetto.) Over time expansionary trends within the economy may require that the migrants be allowed to move upwards somewhat, as long as others are available to fill their positions in the lower strata and their upward movement does not challenge the position of their hosts. Economic assimilation might be more rapid than social assimilation, therefore, because the migrants are necessary to the society's economy but are undesirable as social peers.

This distancing of hosts and migrants, and the associated definition of separate territories, is an extreme example of a more general use of area as a means of expressing independence and reducing conflict. According to some scientists, many animal and bird species have a genetic trait known as territoriality, which involves the need for a home area that can be defined as personal property and defended against all trespassers. (A major use of this home area is during the mating season.) According to the arguments, this trait exists in man also; each family's need for a separate home, the exclusive rights to which are defended in law, reflects a similar territoriality. Thus in mass societies of urbanized nations individuals and households protect their property rights by combining with their neighbours to form a group territory which they control as a community. Thus a hierarchy of territories can be defined, each associated with a particular group: the smallest is the individual household in its own dwelling, and at successively larger scales one has the neighbourhood, the district, the municipality, the state, and so on.

There is a major debate focused on whether territoriality is genetically imprinted in man or whether it is induced by societies in which private ownership of property is a foundation of the economic organization. Whatever the origin, however, there is little doubt that groups in cities do define both their own territories and those of others, and that these are used to separate competitors for the limited amounts of income, wealth, status and power that are available. Thus street gangs in American cities define their 'turfs' with graffiti, for example; Roman Catholics and Protestants retreated from mixed residential areas into segregated ghettos after the onset of the Belfast troubles in 1969; and for many years restrictive covenants were used to bar blacks from a large number of

American residential areas. Territoriality is thus one aspect of the spatial imprint of distancing, a crucial geographical component of the inter-class conflict that characterizes capitalist societies.

Distancing, territoriality and segregation

The discussion so far has been at a general level, outlining the socio-economic rationale for the residential separation of class groups and for the segregation of migrant groups from the host population. As indicated in the introductory discussion, the precision of the theory is not entirely replicated on the ground because, for example, of blurred boundaries between classes (and sometimes between migrants and hosts, as with mixed marriages). Thus although nine classes were identified above, in most cities the number relevant to a discussion of class segregation is probably smaller than that, with the actual number reflecting peculiar characteristics of the local society. Thus the theory presented here provides a framework for analysing individual cities rather than a model which is directly applicable in all. The role of distancing and territoriality as major influences on the social layout of cities is accepted, and the remainder of this chapter provides two illustrations of their operation. These examples are extreme cases, and represent the sort of residential pattern which emerges when one group within society is powerless; the following chapter looks at the more usual patterns, in cities where power is slightly more equally distributed.

Apartheid city

The previous discussion of immigrant group segregation suggested, as is usually the case, that the immigrants so treated are a minority within the urban population and that the segregation resulted from majority group desires. In some situations, however, a minority group with sufficient power has been able to impose its will upon a majority.

Some of the clearest examples of this exercise of minority group power can be found in colonial cities, where a small alien élite dominates and has been able to design, and maintain, the major lineaments of the intra-urban spatial structure. Colonies where the indigenous population was not brought into the capitalist system, such as Australia and the United States, lack this feature: indeed in both of those the near-extinction of the aborigines, largely as a result of harrying by their conquerors, meant that the colonists soon formed a majority. But on the Indian subcontinent the small British contingent was outnumbered by the indigenous population by a very large ratio. The former insisted on residential distancing, defining their own settlements on the edge of the Indian built-up areas (settlements often termed 'The Lines' or, more generally, cantonments).

Throughout Asia, Africa and much of Latin America, decolonization has removed the power of the alien élite with the handing over of economic and social control to the indigenes. The rigid definition of 'white' residential areas has thus been removed, although in many cases those Europeans who have remained have been able, through manipulation of a capitalist property market, to retain their residential separateness. Distancing continues, therefore, sometimes between several groups, as in the cities of Malaysia and Singapore. But one case where the white minority retained its power after decolonization, and has continued to represent its hegemony in the residential patterning of towns and cities, is South Africa. There distancing continues to be dominated by the views of an all-powerful minority group (or at least the most powerful group within that minority – the Afrikaner); the views include a deeply held belief that contact between the various groups creates friction so that a successful multi-racial society requires clear territorial separation of its racial groups in terms of residence, and as far as possible in terms of workplace too.

Racial discrimination has been practised by the minority immigrants since the introduction of white rule to southern Africa, and residential separation has a long history. The whites not only occupy the highest strata of the class system but have also denied access to those strata for the other groups, using a variety of practices, including most recently the reservation of certain occupations, in law, for whites only. Furthermore, members of the other three main racial groups – Indians, blacks (known locally as Bantu) and Cape Coloureds (who have mixed racial origins) – have been excluded from white residential areas, except for those who are employed as domestic servants and who are allowed to live, without any of their kin, on their employers' premises, usually in a small apartment in the back garden. The three non-white groups have also always been segregated from each other, partly for economic reasons (the Indian population, for example, includes many successful entrepreneurs) and partly to avoid inter-group conflict.

This history of social and spatial distancing, designed largely to protect the vested interests of the white upper strata of South Africa's class system, meant that segregation was firmly established in the cities by the middle of the present century. That the members of the lower strata were ethnically different, and thus readily recognizable, simply allowed the discrimination and distancing to be practised very easily and gave the dominant group a basis – perceived racial differences – for justifying their actions.

The pattern of segregation was insufficient for a group within the white population which achieved electoral success for the first time in 1948 and has dominated political life in South Africa ever since. They introduced a much stricter policy of residential distancing as part of their general plan for apartheid, or separate development of the races. The plan recognized that the three non-white groups were necessary as a source of labour power for the country's industries, but determined to keep them apart from their white employers' residential areas. The Group Areas Act of 1950 contained the blueprint for an urban spatial structure. The 'model city' which it defines has been depicted in cartogram form by Davies (Figure 6.1). Its basic components are: (1) members of each group should only live among others of their own race; (2) if at all possible, groups should not live in residential areas which form enclaves, surrounded by one or more other groups; (3) there should be ample room for each group area to expand; and (4) where group areas adjoin, there should be a very clearly delimited boundary or buffer zone separating them. In practice, this means that the city should have a sectoral form, as displayed in Figure 6.1: each group's area may be subdivided on class or other (such as tribal) criteria.

The pattern of residential areas in most South African towns and cities has been altered markedly in the decades since the passage of the Group Areas Act. Figure 6.2 illustrates this for Durban. In 1951 the four groups (there were few Cape Coloureds present) were almost completely segregated from each other, although there were some mixed residential areas near the city centre, but there was a complex mosaic involving much spatial contact of groups along area boundaries. This was altered by the zoning plan introduced under the Act, which did not change the degree of segregation substantially (since this was virtually impossible) but created a much simpler distribution of the separate areas by 1970. The sectoral plan of the ideal apartheid city has been reproduced with great fidelity, and the complex mosaic of mixed and separate areas close to the city centre has been razed. Operation of another aspect of the apartheid programme – the declaration of Bantu homelands with the intention that they proceed to 'independent' nationhood – has enabled almost complete removal of Durban's black population from the city. The homeland of Kwa-Zulu adjoins two sectors of Durban and its boundaries have been so drawn that Durban's two main black residential areas – Kwa-Mashu in the north and Umlazi in the south – are outside the territory of 'white

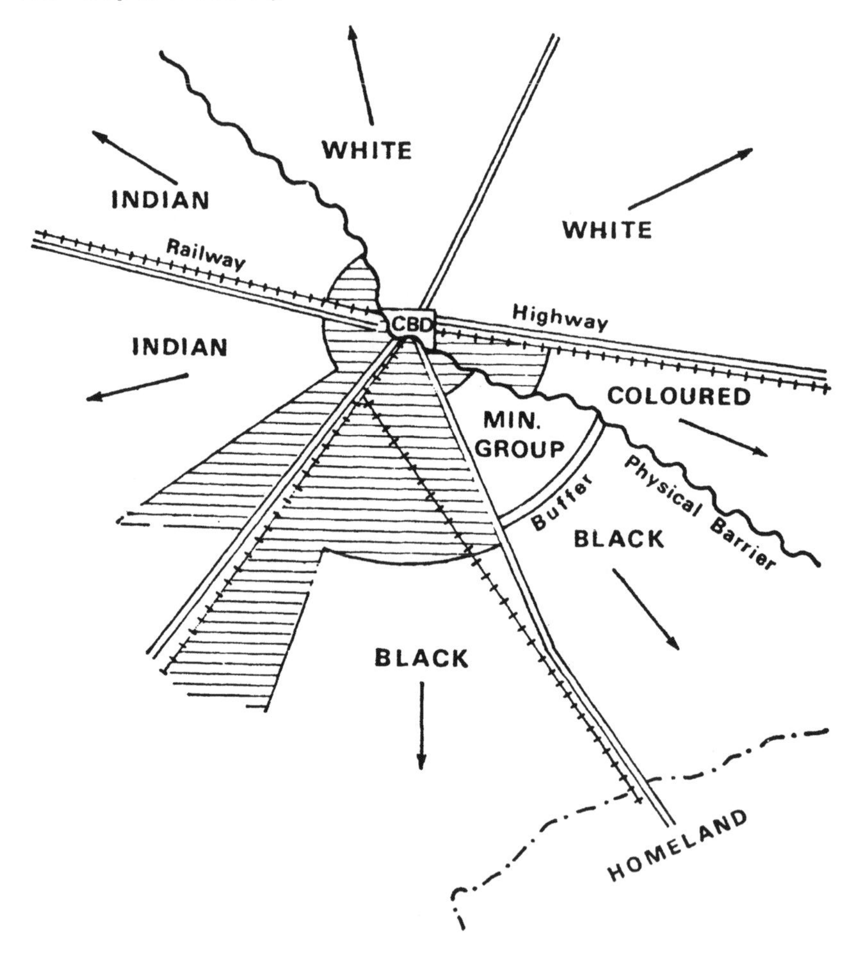

Figure 6.1 The 'model city' of the South African apartheid policy, as drawn by Davies, 1972, p. 803. The shading indicates non-residential areas; CBD= central business district.

South Africa'. If Kwa-Zulu does opt for South-African-defined independence, Durban's black workers will become daily international commuters.

Similar schemes expressing the Afrikaner view of race relations have been imposed in all towns and cities. Thus, for example, Cape Coloureds who formerly lived in a number of cosmopolitan districts close to the centre of Cape Town have been moved to suburban locations twenty miles or more away, where many have been constrained by the housing market to erecting squatter settlement homes. (These settlements are

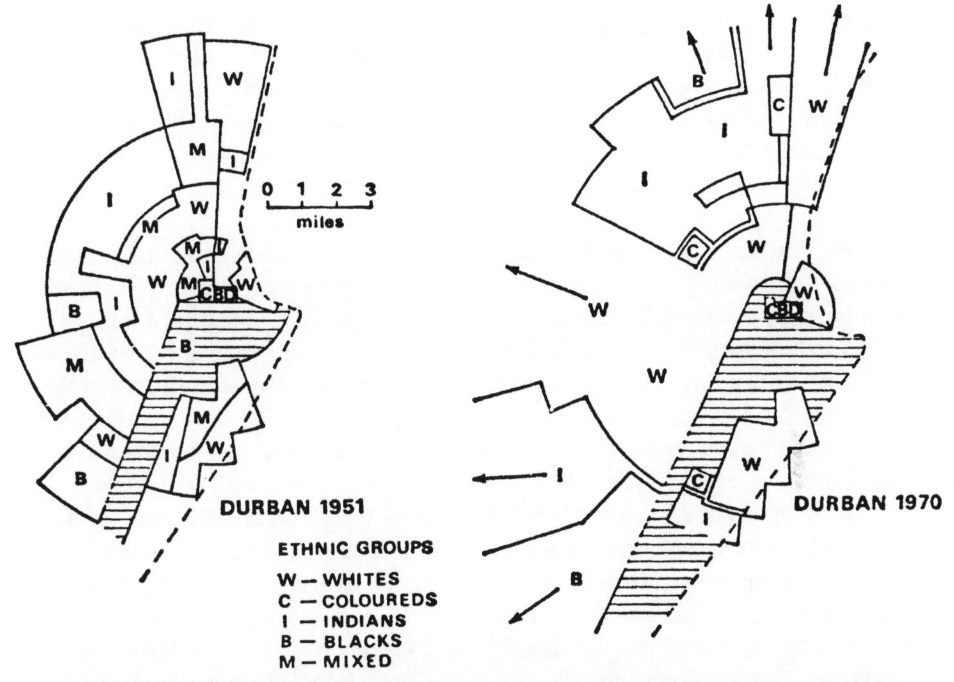

Figure 6.2 The residential pattern of Durban in 1951 (before the Group Areas Act was enforced) and in 1970 (following enforcement of the Act). The shading indicates non-residential areas and the dashed line indicates the coast; CBD =central business district. (Source: Davies, 1972, p. 803.)

legally recognized; those occupied nearby by blacks are not.) Similarly Johannesburg's large black population has been decanted to the suburban ghetto of Soweto. (Johannesburg is not adjacent to a homeland boundary. Allocation of a home in Soweto is dependent on long service as a worker in the city. Initially immigrants must leave their families in their homeland and live in single-men's quarters; only after a decade or more of service can they invite their families to join them.) Finally, Pretoria is about 40 kilometres from the nearest homeland – Bophutatswana – and black workers in the capital are being forced in increasing numbers to commute daily from the rapidly mushrooming squatter settlements just inside the homeland borders.

Municipal exclusion practices in the United States

The residential pattern of contemporary South African cities has been possible because a minority of the population has been able to deny the majority non-white groups any political representation and has employed its political hegemony to ensure its economic superiority and security (backed, of course, by stringent police powers). Such extreme imprinting of the views of the upper strata has not been possible in most other capitalist countries, despite similar views. Thus in the United States, for example, the capitalist cause has been advanced not by the total suppression of those in the lower strata, most particularly the blacks, but rather by their slow incorporation into full economic and social membership of the society: their support has been bought, usually grudgingly, by the granting of democratic rights, by the inducements of potential economic and social mobility, and by the provision of improved living standards. But social separateness has still been the general desire of the higher-class white population, and a variety of mechanisms has been applied in attempts to distance blacks from whites since the abolition of slavery in the 1860s.

Many of these mechanisms have involved legalistic procedures, some of which have later been invalidated. Thus in Louisville, Kentucky, at the beginning of this century city ordinances existed which established exclusive residential areas for blacks and whites. Such apartheid was deemed unconstitutional by the Supreme Court, however, in that it violated the equal protection clause of the Fourteenth Amendment. And so urban whites sought to achieve similar ends without using the legislative process. Individual agreements were promulgated, either in the form of restrictive covenants which were part of the title deed of a dwelling and involved phrases such as 'Subject to the covenant that said lots shall never be rented, leased, sold, transferred or conveyed unto any negro or colored person', or as agreements made by local property owners not to sell or rent properties to Negroes: there were seventy-eight local associations with such agreements in Chicago in 1944. Such practices were also ruled illegal, however, when enforced by courts. Only one practice has so far stood the test of time: suburbanization, accompanied by municipal protectionism.

Suburbanization in the United States during the nineteenth century involved the relatively rich and powerful moving away from the city centres to the pleasanter environments of the rural–urban fringe. Their migration was facilitated by the developments in public transport, and their relatively long journeys to work made feasible by a situation which did not apply to members of the lower strata; the latter were prevented from moving to suburbs by a combination of such factors as their longer working hours, the relative insecurity of their jobs, the high costs of public transport and the timetabling of services. Thus the rich were able to escape what they perceived to be the growing negative externalities of central-city neighbourhoods, such as congestion, various forms of pollution, health hazards and social contact with undesirable lower-class groups.

In the new suburbs the residents were content to live under the political jurisdiction of the city and to contribute to its government, and they consented to extensions of the city boundaries so as to incorporate the new residential developments. Late in the nineteenth century, however, this attitude changed. Instead of accepting city annexation of the new residential areas, the suburbanites opposed it and incorporated their own independent municipalities, under the relevant state constitution, to provide them with separate local government. This generated much conflict between the city and suburban governments, often over the supply of facilities such as water to the latter. In some cases the suburban municipalities were forced to yield to pressures for annexation, but most stood firm. The result was a balkanization of suburbia, which exists still and has been expanded manyfold.

Suburban political separateness was imposed by its residents for a variety of reasons. Firstly, the higher classes found that their power in the central cities was being diluted by the spread of the franchise and the politicization of the lower classes. By insisting on political separateness the suburbanites were able to obtain control of their home neighbourhoods, even if they had to yield control over their workplaces. Secondly, steady expansion of the role and functions of municipal governments meant increased demands on local taxes (raised on the basis of property values) to finance the various programmes which benefited the relatively poor but whose costs were disproportionately met by the relatively affluent. By forming their own separate municipalities, the rich were able to avoid this subsidizing of the poor through the local tax system, whilst still enjoying many of the facilities provided by the central-city governments and paid for by their residents.

The third reason is concerned with the growing prosperity of urban Americans over time, which provided an increasing proportion of them with the means of escaping the inner-city externalities and moving to the pleasanter suburban environments. This threatened the exclusive character of the existing suburbs, and challenged the distancing mechanism which had been established there. Incorporation of separate municipalities offered an answer to this challenge (although some of the wealthy had to move further out in order to activate it) because in most states the municipalities have widespread independent zoning powers under town planning legislation. The class in power in each municipality could zone for a land-use pattern that would exclude those developments considered undesirable, notably industry and low-cost, high-density housing which would be occupied by the lower classes, including blacks. Use of the zoning powers allowed the residents of each municipality to create their own ghettos, excluding those with whom they did not want to share their neighbourhoods. Thus the outer edges of most American urban areas are administered politically by a complex mosaic of small municipalities, each operating policies which favour the interests of the small resident population only. These may not all be of high socio-economic status, for a variety of upper- and middle-class, and a few lower-class suburbs have been established; indeed there are suburbs zoned by industrialists to keep people out, and so allow the firms owning the local factories to avoid making any contribution to the costs of providing the sorts of services (streets, water supply, etc.) which their workers require of their local governments.

In most states a further important advantage offered by suburban areas is independent control of the local school system. Public education is organized in the United States by *ad hoc* school districts, of which there are many thousands; these are independent of other bodies in their powers over such issues as teacher numbers and salaries, expenditure on facilities and syllabuses. The territory of the central city in an urban area will have a single school district governing all institutions in its area, but the suburbs are characterized by a large number of small districts serving either only one municipality or a few adjacent ones. Thus a combination of land-use zoning powers and control of local schools allows the residents of suburban municipalities to ensure that the education provided for their children, and the pupils with whom they share it, are consonant with their educational aspirations for the next generation. Further, while at the same time excluding perceived negative externalities from their children's schools, the suburbanites can avoid having to pay through local taxes for the education of those poorer children, most of whom will be in receipt of an inferior education in the inferior facilities of the inner city.

These various aspects of suburban independence have been of great importance to local residents in recent decades. In 1954 the Supreme Court ruled that the provision of separate schools for black and white, albeit with equal facilities, was unconstitutional under the Fourteenth Amendment. Schools were to be racially integrated, with approximately the same racial proportions being produced for each school in each district; busing was the approved means of achieving the integration. But as the suburban school districts contained very few black pupils and the busing did not involve any inter-district transfers of pupils, the suburban school districts stood out as fastnesses for whites, especially the more prosperous whites, where they could avoid the negative externality of their children being forcibly sent to the same school as black students. Fears that this exclusiveness might be threatened by a centralized control of education were removed in the 1970s when the Supreme Court ruled that the uneven distribution of resources (local taxes) for education was not in violation of the Constitution.

The use of discriminatory land-use zoning to ensure suburban exclusiveness has been challenged recently before the country's courts on a variety of grounds, including the equal protection provisions of the Fourteenth Amendment. A few cases have been successful: in New Jersey, for example, municipalities must now zone for a variety of housing types and costs (though there is no mechanism to ensure that this variety is achieved). But in many states the zoning powers of municipalities have been upheld, in some with the support of local referenda on particular issues, and suburbanites have retained their rights to determine who can afford to live among them.

Suburban exclusiveness has been sought in many parts of the world. In some nineteenth-century British cities, for example, high-class suburban developments were walled in and their entrances protected by policed gates; similar estates are still being developed by private companies, less so in Britain than elsewhere. (It was only a few decades ago in Oxford that the residents of a middle-class neighbourhood built a wall across a street to distance themselves from the people of an adjacent public housing estate.) The separate municipality ploy has also been attempted widely. Many large British cities were ringed by a zone of small urban districts, for example. These semi-independent units were dissolved with the local government reorganization that came into operation in 1974 and provided authorities which administer the entire urban area, a solution often proposed but never fully instituted in the United States where suburban independence continues despite its many challenges.

Conclusions

These two examples illustrate how the principles of distancing and territoriality can be operated by a small but extremely powerful élite group to manipulate urban residential patterns, thereby protecting their interests in terms of access to positive externalities, escape from negative externalities, protection of property values, and control of the educational system which allocates income, status and power to the next generation. The apartheid and municipal solutions to the requirements of the upper strata of the class system for maintenance of their class positions in a highly competitive, rapidly changing economic system are extreme solutions and are relatively rare. Where the upper strata lack hegemony, or even dominance, over the members of the lower classes they must find alternative ways of preserving their positions. These ways usually involve manipulation of the property market, as will be discussed in the next chapter.

7. Housing Classes, Housing Markets and Urban Residential Patterns

In the previous chapter a series of social and economic factors was outlined as the reason why various classes and groups within a society wish to distance themselves from other groups and classes, the members of whom are perceived as threats to the established economic and social order. The present chapter investigates how these desires are operated in societies where the dominant groups lack the total hegemony described at the end of the last chapter. First, however, a brief description of the general pattern of segregation in capitalist cities is provided, as a backdrop to the later explanation.

Contemporary residential segregation

The facts of residential segregation are very obvious in virtually all cities, to both their residents and outside observers. In the large cities there may be many districts and neighbourhoods which are *terra incognita* to large proportions of the inhabitants, who have no reason to visit them or to learn about them. Nevertheless, most urban residents have a general mental map of their home town which delineates the major features of its social geography and agrees in overall form if not in detail with the similar map held by each of their contemporaries. Most have information about the areas at the extremes of the social scale: the most 'fashionable' areas are usually recognized, as are the 'worst slums', even if neither has ever been visited. Specific elements of the mental maps refer to the area around each individual's home. The local neighbourhood provides the activity space within which much of the resident's social life, and perhaps economic life too, is enacted. Thus within the generally agreed mental map of the city's social topography are area-specific elements reflecting the particular knowledge of the residents of each area.

These mental images provide the structuring framework within which urban inhabitants organize their daily lives; understanding the content of these images allows detailed investigation of how people behave within cities – where they shop, who they meet, where they would like to live, and so on. For the study of the entire social map of the city, however, the researcher usually requires some more easily manipulated data. These have been provided in recent decades by census authorities, who collect and circulate a great mass of information on the characteristics of people, households and housing. Such information is collated for small areas within cities – in some countries for areas with populations fewer than 1,000, so that very fine-grained analyses are possible. This information has been widely used by urban geographers and researchers in cognate fields.

A simple method for using these census data sets to display the degree of spatial separation of class and other groups is illustrated in Table 7.1. The data there are for a hypothetical city divided into eight districts, whose populations are divided into four social classes (A to D). How separate are A and B in their residential distributions? This can be assessed by taking, separately for each district, the difference between the percentage of those in group A living there and the percentage in group B living there. This is the row $|A\% - B\%|$ in the table, with the vertical lines indicating that the difference is recorded irrespective of sign. These differences are summed over all districts, as shown in the table, and the sum is divided by two to provide an index of dissimilarity (for A and B this is 78·8/2=39·4). The index can range from 0·0 (indicating no dissimilarity) to 100·0 (indicating complete dissimilarity). The larger its value, the greater the residential separation of the two groups; a more detailed interpretation is that the index shows the percentage of the members of one of the groups who would have to be reallocated to different districts so that its distribution (in percentage terms) was the same as that of the other. The full set of indices for the hypothetical city is given at the foot of Table 7.1, and indicates that classes A and B tend to occupy the same districts (as also do C and D) but that they are not mixed with the members of the other two groups.

A second question which might be asked of the data in Table 7.1 is 'How separated is one group from all of the other groups?' (or 'To what extent do the members of any one group live apart from all of the others?'). This can be answered by the index of segregation, which is similar to the index of dissimilarity; it is computed as half of the total difference between the percentage distribution for the group – in this case A – and the percentage distribution for all of the other groups – in this case (B+C+D). This is shown in Table 7.1 in the row $|A\% - (B+C+D)\%|$. The resulting index of segregation is 63·3, which indicates in percentage terms the degree to which group A is residentially separated from the rest of the population.

Table 7.1. The index of dissimilarity

District	*1*	*2*	*3*	*4*	*5*	*6*	*7*	*8*	*Total*
Social Class									
A	80	60	40	10	5	3	1	1	200
(%)	(40)	(30)	(20)	(5)	(2·5)	(1·5)	(0·5)	(0·5)	(100)
B	20	20	40	20	10	10	5	5	130
(%)	(15·4)	(15·4)	(30·8)	(15·8)	(7·7)	(7·7)	(3·8)	(3·8)	(100)
C	5	5	10	10	20	50	50	30	180
(%)	(2·8)	(2·8)	(5·6)	(5·6)	(11·1)	(27·8)	(27·8)	(16·7)	(100)
D	1	1	3	5	10	10	15	40	85
(%)	(1·2)	(1·2)	(3·5)	(5·9)	(11·8)	(11·8)	(17·6)	(47·1)	(100)
|A%−B%|	24·6	14·6	10·8	10·8	5·2	6·2	3·3	3·3	78·8
B+C+D	26	26	53	35	40	70	70	75	395
(%)	(6·6)	(6·6)	(13·4)	(8·9)	(10·1)	(17·7)	(17·7)	(19·0)	(100·0)
|A%−(B+C+D)%|	33·4	23·4	6·4	3·9	7·6	16·2	17·2	18·5	126·6

Indices of Dissimilarity

A−B 39·4 A−C 78·9 A−D 84·2

B−C 60·5 B−D 65·5 C−D 31·5

Indices of Segregation

A 63·3 B 31·0 C 55·6 D 51·5

These two simple measures have been used to assess the extent of residential separation in a large number of cities. In general terms, the results for social class groups show that:

(1) the greater the social distance between two classes (assuming a rank-ordering on a scale such as that on p. 154), the greater their index of dissimilarity so that, for example, semi-skilled and skilled workers might be mixed together in the same districts but semi-skilled and professional workers are very unlikely to be; and

(2) the index of segregation is greatest at the extremes of the social class scale so that it is those with the greatest amounts of income, status and power and also those with the least who are most isolated spatially from members of other classes.

Similar findings have been reported for income groups from cities in many parts of the world.

Much higher indices of dissimilarity than reported for class separation have been identified in studies of various minority groups, mostly racial and religious but in some cities cultural/linguistic (as with the French and the English in Québec). In 1960 for example, the average index of segregation for blacks in 109 United States' cities was 86·1, and there is no evidence that it has been reduced markedly since. Such separation is much greater than could be anticipated on the basis of class and income differences alone, and experiments have suggested that in one city the index of segregation for blacks was fifteen times higher than would be the case if income were the only factor influencing residential location. Further, it is almost certain that such indices understate the degree of segregation. They are based on areas not specifically defined for the purpose of measuring segregation and whose boundaries may split a clearly defined neighbourhood into two or more separate districts; all measures of segregation are only as good as the data sets on which they rely.

During the last decade or so, researchers have been developing a series of more sophisticated measurement procedures, which are commonly known as social area analysis and factorial ecology. These analyse large numbers of distributions simultaneously, and allow statements to be made about the general patterns of residential differentiation for many components of a city's population. The results suggest the existence of three main types of distancing mechanism within urban areas, though this conclusion is entirely dependent on the nature of the data collected and made available in censuses. The three are:

(1) Separation of the main social classes within society, as indicated by data on such features as occupation, income, educational attainment and property values;

(2) Separation of minority groups, independent of their separation on class grounds, from their host population and from each other. In many cases the areas occupied by a particular minority are themselves spatially structured on class grounds with, for example, low-income blacks in American ghettos living in separate districts from the high-income blacks; and

(3) Separation of different types of household, which comprises two interdependent patterns. The first identifies the spatial separation of 'conventional' households comprising nuclear families, living in their own separate dwellings, from non-family households, many of which rent apartments. The other isolates differences among the nuclear family areas in terms of population age and sex structures – young families tend to live apart from those in which the children are teenagers, who in turn are separated from those whose children have left home.

With minor variations, these three patterns are repeated in the towns and cities in most of the capitalist world. In small towns, and in the cities of some countries of the Third World, the degree and type of separation is sometimes different, but the generality of the findings summarized above is one of the major recurring themes of urban geographical research in recent decades.

Housing provision and housing classes

The search for an explanation of the patterns just described follows the arguments of the previous chapter. Members of each class wish to distance themselves from those whom they perceive to be lower on the social scale, because of the negative externalities that residential mixing with such people is thought to bring. In capitalist societies they have achieved this separation by making land and housing into commodities which are traded in the same general way as are other goods and services, with price reflecting the interaction of supply and demand. In simple terms, this would suggest that the highest-income groups get first choice of areas, since they can outbid all others; the next most affluent get second choice; and so on to the lowest-income group, who get no choice at all. To achieve this solution, however, a complex set of economic and social mechanisms is required.

Property values

For most commodities, the price that consumers are prepared to pay reflects their utility for consumption: the good or service has a use value to them, and if the price exceeds that value, they forego its consumption and purchase something else instead. Property (an inclusive term for both a home and the land on which it stands) differs from some other commodities in that it must be purchased, since each family must have a home and this requirement cannot be substituted by other goods; the only possible substitution refers to the size, type and location of the home. It also has some other relatively peculiar characteristics. Like some goods – a Rolls-Royce car, perhaps – it has a status value. Property is a positional good, indicative both by its nature and by its location of its owner's or occupant's status in society. For those who wish to display their status, and can afford to spend more than is absolutely necessary for an adequate shelter, the use value of their property reflects more than its worth to them as a machine for living.

Three other important characteristics distinguish property from most other commodities:

(1) The land component is permanent, although damage is possible, so that it can be traded after several decades of use in the same condition as when it was bought (this refers to urban land only, of course, for rural land can be raped by poor husbandry);

(2) The supply of land is virtually fixed, unlike that of many other commodities, so that the seller is usually in a strong position relative to the buyer; and

(3) Dwellings have a much longer average life-expectancy than is the case with most consumer goods.

Because it has these characteristics, property is an investment as well as a commodity to be consumed. As a consequence, people may be prepared to pay more than a price consonant with the use they expect to get from a property over a given period; the extra can be interpreted as a premium paid to secure ownership of something whose value in the market place might be enhanced with time (relative, that is, to the general course of inflation). Indeed, it is because of the fixed supply of land in a town that property is considered an excellent investment by many. (Land supply can be increased slightly by higher-intensity usage, and towns do expand in area; with the latter, however, without a spatial restructuring of other land uses the new area on the edge of the city is relatively disadvantaged with respect to accessibility to the city centre.) To obtain such investments, therefore, particularly perceived good investments with promising future positive externalities, people may be prepared to pay relatively large sums, even borrowing money for this if the expected return is greater than the rate of interest on the loan. Thus property is now a major commodity in capitalist societies, invested in by financial institutions and speculators as well as urban residents. The property market is part of the capitalist operation.

Housing market evolution

The evolution of the urban housing market can be traced from the mercantile settlements, in which only a small proportion of the households owned properties. Many employees, especially the unmarried ones, lived in their employers' homes; some rented their own, from landlords who were either mercantile capitalists or rural landholders and who had invested in providing housing for urban dwellers. Payment of rent required a regular income, and many families ensured that they could afford this by taking in lodgers.

With the introduction of industrial capitalism, some employers found it incumbent upon them to provide housing for their employees, either for rent or free (the latter in lieu of part of their wages). Some provided barracks for the workers only; others built cottages for workers' families, with the cottages varying in quality, size, and sometimes location too, according to the occupants' employment status. But such provision of housing restricted capitalist activity, for it tied up some of the industrialists' capital which might otherwise be used in the search for profits; it also created difficulties for employees, who were somewhat restricted from seeking to move to other employers. (This latter was a constraint on competing employers, too.) Thus as towns grew and developed into competitive labour markets rather than locales monopolized by a single employer, so the provision of housing was separated from the provision of work. Entrepreneurs perceived that investment in the former might be a profitable use of capital. Construction of workers' dwellings, to be rented rather than purchased, was undertaken in the expanding towns, therefore; many small firms were involved, building and renting only a few homes each, but there were also companies which bought larger tracts of land and constructed housing estates such as the brick terraces of nineteenth-century Britain.

This first incursion of capital into the urban housing market was undertaken solely for capitalist ends – an acceptable profit margin – and the size and quality of the homes were tailored to meet the incomes of the relevant classes. Thus, for example, 'better' homes were built for the lower white-collar classes than for their upper blue-collar contemporaries. (Many of the former have always been prepared to pay a higher proportion of their incomes on housing than have the latter, in order to attain the perceived positive externalities of their status.) But the higher-status white-collar workers, the professional and managerial classes, were provided for in the separate, ownership market.

Power and the ownership of land were closely interlinked in pre-industrial societies. It was the landowners who ruled and were represented in parliament, and as the franchise was slowly extended it was the small property-owners of the towns and cities who benefited first. As property-ownership represented power and status, it was natural for those who aspired to elevated positions in the emerging urban–industrial society (some of whom had roots in the earlier landowning classes) to seek to be owners themselves. Thus land was yielded by agriculturalists, at prices exceeding those reflecting the value of the land for farming activities, on which custom-built homes could be provided for the *nouveaux riches*. Small building firms undertook the construction, and financiers were prepared to advance money to those with substantial incomes as mortgages to be paid back over long periods of time at rates of interest which gave acceptable returns on the capital invested.

In its initial stages, therefore, industrial capitalist urban society was divided into a relatively small, property-owning white-collar class (the petty bourgeoisie), an even smaller urban capitalist class who owned large properties, and a large proletariat who rented their homes. This division has been changed with the movement towards late capitalism. In particular, the rental sector has declined, in part because of the relative fall in the size of the proletariat and the associated growth of the petty bourgeoisie, reflecting the alterations in the division of labour. A variety of other reasons accounts for the growth of property-owning, however, of which four are of major importance.

(1) The provision of rental housing has turned out not to be a very profitable venture for capitalists seeking profits. As dwellings age, so they need to be maintained, and unless the costs can be passed on to the tenant the landlord must meet them out of his potential profits. High rents, especially if they apply to poor-quality dwellings, may be resisted by low-paid tenants, however, producing a limit to the potential for increases. Thus although the provision of housing offered a reasonable and safe return for landlord-investors, competing opportunities for capital offered better chances of 'getting rich quick'.

(2) The status involved in home ownership became widely recognized and desired. (The power declined with the granting of universal adult franchise, although in New Zealand counties until the early 1970s property-owners still had three votes in local government elections whereas all other residents had only one.) The ethos that developed was that owning one's home reflected status in society; in some societies, such as

the American and Australian, this was part of a more general individualistic ethos based on the concept of landownership. In addition, ownership represented an investment; rent is money spent on obtaining the use of a commodity but a mortgage repayment is an investment in an equity, in a particular commodity that can be sold, very often at a higher price than it was purchased for. Encouragement of home ownership as a respectable and desirable aspiration among the proletariat, as well as the petty bourgeoisie, allowed the advancement of capitalist desires, for two reasons. First, because of the equity component in mortgage repayments which is absent from rents, households can be persuaded to pay more for the former than for the latter, thereby offering greater (as well as more rapid) returns to those who invest in the production of housing for sale than to those building homes to rent. Secondly, ownership defuses some of the latent conflict in society between the forces of capital and labour. Home ownership, it is widely argued from the capitalist standpoint, stabilizes society; it reduces the potential radical demands of those who have a small investment to protect, and produces a conservative, home-owning class. (One proposal after the 1976 riots in Soweto township outside Johannesburg was that if blacks were allowed to own their homes they would be much less prepared to attack property.)

(3) Home ownership assists the circulation of capital. People who have been convinced of the desirability of ownership can be persuaded to save money towards the down-payment that is needed for a purchase, and then to pay the interest to those who lend money for mortgages. These savings can be invested by the financial institutions they are lodged with, thus increasing the proportion of the society's income which is devoted to the search for capitalist profits (some part of which is returned to the savers). Thus saving and the associated home purchase become parts of the ethos of capitalist society, replacing current consumption by the promise of greater benefits tomorrow.

(4) Ownership assists in ensuring the profitability of the construction industry. One of the problems of late capitalism, as indicated in Chapter 2, is guaranteeing sufficient demand to absorb the output of its increased productive capacity. Home ownership helps to maintain demand for certain commodities, especially if people can be convinced that they should always be aspiring to a better, often a bigger, home than the one they currently occupy. Many segments of industry are supported by the new housing requirements, involving not only the building firms but also those who provide roads, public utilities, the cars that are needed to reach the new homes, the various hardware that makes up the contents of a 'modern' home, and so on. Some potential problems of over-

supply can be countered by encouragement of the desire for a new home, therefore.

For these four interrelated reasons, home ownership has become a major household goal in most late capitalist societies; it provides independence, a small, hopefully appreciating, investment, and a status symbol for the owners, while it also stimulates necessary demand that maintains business viability. Part of the operation of one sector of capitalism is oriented towards supporting and protecting this aspect of society, therefore, and it is reflected in the political ethos – a 'property-owning democracy' – of many countries.

Housing market operations

Producers of commodities in capitalist societies have a continuing problem of balancing supply against demand, for if the latter falls relative to the former then prices are likely to drop, the extent of the drop reflecting the elasticity of demand for the commodity. This problem faces housing-providers, who have sought to ensure the profitability of their businesses. Demand for housing must be maintained, and several means are available to ensure this.

(1) Maintenance of a buoyant market at all times, which is best achieved by convincing consumers that they should constantly be aspiring to better, newer housing. The major targets of such propaganda are usually the most affluent classes, because new housing for them is likely to be relatively expensive and so profitable for the builder; the richer households in general spend relatively less of their income on housing so they are best able to meet increased costs. The lower the class for whom a housing development is aimed, the greater the speculation that the members will be able to afford the price which will ensure an acceptable profit for the builder.

(2) Establishment of a monopoly, with the attendant pricing benefits. In theory this is feasible, since one of the main elements – land – is in relatively fixed supply for the population of any one town or city, and housing is a necessity. But demand is not inelastic, and if housing becomes too expensive people will purchase relatively small amounts, and invest their money elsewhere. Further, ownership of land around urban settlements is usually so dispersed, and the construction industry divided into so many small firms that, despite much concentration and centralization in the industry, the monopolistic stage is not near.

(3) Maintenance of a chronically poorly housed sector of the population, which ensures a continued demand; the poor housing standards

may only be relative, but should be such that the demand for better is sustained. If supply does not meet demand, density of occupation will increase in the areas of poorest-quality housing, and this may create negative externalities for the residents of adjacent areas, who fear a spilling-over of the conditions into their own neighbourhoods and wish to be distanced from them. Their desire to move away fuels the demand for new homes at acceptable prices and profit levels for the builders.

(4) Encouragement of new demand. This cannot be undertaken directly by the construction industry, although it can be connived at. For example, a trend towards younger age at marriage, associated with a desire for a separate home immediately and perhaps with children leaving the parental home and establishing separate households prior to marriage, increases the demand for housing relative to the rate of population growth.

All of these solutions are operating in capitalist societies. The fourth is currently very noticeable in many countries, for example, but in the long term it is probably the first that offers the greatest potential returns. Indeed, it is part of a general theory of housing provision in capitalist societies known as filtering. This states that every new home occupied by an affluent household not only provides profits for a builder but also leads to the vacating of a home which can be occupied by a member of a lower-income group. In turn, this produces another vacancy, and so on, so that one new, expensive home stimulates a chain of moves. At each step the household concerned is able to move into something better so that, it is argued, all classes benefit from the provision of new housing for the rich (a result which would not be brought about if the new housing were built for lower-income groups). But critics of such a policy point out that it does not work to the benefit of all: the upper classes are relatively small, and could not move frequently enough to ensure a substantial supply of second-hand homes for all those not being catered for by new construction.

Because most new housing is built in capitalist cities for the more affluent residents, and the filtering mechanism does not work to provide better housing for all, almost every place has a chronically poorly housed sector. Their relative suffering reflects the interaction of a variety of causes. Most important is their low and often irregular income, which prevents saving and makes it difficult to afford rents. Sharing of dwellings by several households is often the only solution, since it allows the pooling of resources to meet rent demands, but this almost certainly leads to poor conditions and discourages landlords from spending on maintenance. The consequence is that the poor are confined to the worst housing – the classic slums.

A solution adopted by the poor in many cities, especially those of the Third World where the problems of low incomes and irregular employment are most acute, is the provision of their own housing illegally. These are the squatters, the residents of shantytowns who occupy land illegally and build their own homes on it, using whatever materials they can obtain and extending and improving the buildings whenever this is feasible. The stereotype of these areas is that they contain the radical elements within the proletariat, living in unhealthy environments and creating many problems, political and otherwise, for the city authorities.

The poor health conditions in these squatter settlements are undoubted, although they are frequently no worse than the densely occupied rental housing of the inner-city slums. But in most cases their residents are far from being dangerous radicals. Two types of settlement can be identified. The first is the slum of despair, housing what are sometimes known as the toeholders in society, those living on the fringe of the urban economy with neither permanent income nor any hope of it. For them, regular rent payments are impossible and building a shack on a vacant piece of land is the only way in which they can obtain housing. The other type, the more common in most cities, is the slum of hope, occupied by what are known as the bridgeheaders, those who have established themselves in the urban economy. They have regular, although small, incomes, but by living in squatter settlements, and therefore paying no rents, they avoid devoting a large portion of that income to the cost of low-quality housing. The squatter settlement housing is an investment in a small piece of equity (even though it may not be realizable); more importantly, it allows its residents to avoid some of the worst aspects of the rental market, and they can retain most of their income for the purchase of other needed goods and services, including the provision of a home base for their children's economic and social mobility. Occupants of such squatter settlements are thus in many cases part of the 'conservative, home-owning sector', who organize their settlements in order to advance their social and economic aspirations. Indeed, in many cities some of them are members of the white-collar classes who have found that the self-help housing of the settlements offers the most realistic positive externalities, because the conventional housing market can never cater for the volume of demand in such low-income societies.

Squatting is not a characteristic of Third World cities only, although it is most prevalent there. Inability to afford even the lowest rents leads to some homelessness in most cities. In London recently, for example, it has stimulated squatting in empty dwellings (some of them vacant during a change in ownership but most while awaiting demolition), and in Cape Town the removal of Cape Coloureds from the inner-city areas, and the provision of public housing for only a few low-income families, has led to the development of suburban squatter settlements among the sand-hills. In both of these situations, however, the size of the squatting 'problem' (or 'solution', depending on the point of view) is small relative to most Third World cities, and it has been more rigorously countered by the powerful groups in society. (Squatter settlements have been energetically repressed in Third World cities, too, but usually with little success; they are now widely accepted and tolerated.)

The state and the provision of housing

In many societies there is insufficient housing of acceptable quality relative to the demand for it, and coping with this problem is a task which has been accepted by governments, often with considerable reluctance. In part they do this as one element in their attempt to defuse the latent conflict between labour and capital; in part they may do it because they have been elected to represent the needs of the working class within the capitalist system; and in part they do it to protect the interests of the construction industry. For these and other reasons governments have initiated a wide range of housing policies.

By far the most obvious government housing activity in many countries, though not necessarily the most important, is the provision of public housing. Four types of public housing can be identified. The first involves urban renewal – often termed slum clearance – which aims to provide better-quality homes for the poor, most of whom live as tenants in older dwellings close to city centres. This kind of public housing dominates in the United States, for example, where legislation requires city governments to replace each slum dwelling demolished for health or other reasons by at least one unit of public housing. It is now the largest component of the British council housing programme too. In both countries most of the new housing is provided in inner-city areas, where the old inadequate dwellings have been demolished. Here the difficulties of amassing sizeable areas of land and of combating high land values, and the lack of alternative sites for the new homes (because the whole of the territory of the relevant local government unit is built over) have led many governments to build tower blocks of flats as the main elements in urban renewal schemes. In fact, the overall densities produced have usually been no greater than those of the low-rise dwellings being replaced, because of the need to provide open space; as discussed below, the environment of these tower blocks has not been widely acclaimed.

The second type of public housing involves meeting shortages of supply through subsidized suburbanization. For several decades this was a major component of the British public housing programme. Local governments either build the homes themselves or contract with private firms to build large estates of single-family homes in suburban locations to cater for those, again many of them tenants of poorer-quality dwellings in the city centre, who cannot afford homes available for sale from private builders. Thus the housing provided in these schemes has generally been of a high quality, relative to the rents charged and the costs of comparable housing in the private sector. In Britain most of this housing has been built by individual local governments within their own territories, to meet the needs of local inhabitants only. In some cases, however, a lack of available land has led authorities to contract for development in the territories of other local governments, as with some of the large estates, such as Dagenham, built by London authorities.

The third type of public housing has been used to stimulate development. Such a policy may have been a general one, such as after a war or a major economic depression which has left a major housing shortage as a residue. More usually it reflects the need to stimulate growth in a particular place. In Britain the obvious examples are the new towns established after the Second World War to provide a new form of suburbanization beyond London's green belt and the town development schemes in places like Basingstoke and Swindon which were introduced as part of a policy of dispersing population and economic activity from the metropolis.

Finally there is the public housing which caters for the needs of particular disadvantaged groups. Preponderant among these are the elderly. Many of the older people now benefiting from medical advances and the possibility of retiring relatively early are neither physically nor financially able to cope with a conventional property. Thus smaller dwellings, often in communities and in some cases with a social worker attached, have been designed for the elderly; one quarter of all public housing currently built in the United States is for this group.

The provision of public housing is opposed by certain groups in capitalist societies, because of its obvious subsidization and its opposition to the general ethos of ownership. (Some governments face the latter opposition by arranging for the sale of public housing to tenants.) Most governments, too, prefer the development of a 'property-owning democracy' to the enlargement of the 'state tenancy', and have designed a variety of policies and initiatives aimed at ensuring that homes are built for, and can be afforded by, the majority of households.

Many of these policies are fiscal. In New Zealand, for example, which has one of the highest home-ownership rates of the advanced capitalist nations, interest rates on mortgage loans in the private sector have been kept low (until the recent rapid inflation) by a financial policy which manipulates the volume of money rather than its cost. In addition, the government itself operates a large mortgage fund, with interest rates below those of the private sector (and even lower for those with incomes below a certain threshold). It also has a scheme whereby the weekly child benefit for families can be taken as a lump sum to use as the deposit on a home.

In Britain, and in several other countries, interest payments on mortgages are eligible for tax relief, as are the insurance premiums on policies used as collateral for the mortgages; the interest on deposits in building societies, whose prime function is to advance mortgages, is also allowed tax relief. In the United States the Federal Housing Administration has been empowered since the 1930s to insure mortgages against their recipients' default, thus protecting the lenders' risk and allowing them both to reduce their interest rates and to extend the period of the loan. Another federal agency redirects money into the mortgage market from other sectors of the capital market when the supply of the former is low because higher potential returns on investment are available elsewhere; the result is a federal subsidy for mortgages. And so the list could continue. In almost every capitalist country governments have used fiscal instruments to encourage home-buying, aiming not only to ensure a 'decent home' for every citizen but also to win consensus support for the state and for the political party in power.

In cities lacking a large squatter settlement population, the third major component of the housing market, after private ownership and public rental, is the private rental sector. For these tenants some governments have tried to ensure both that rents are not too high relative to incomes and that the dwelling quality is adequate. In addition, some have also legislated to give tenants reasonable security of tenure. The problem with fair-rents legislation, with its limits to landlord action, is that it can easily both discourage further investment in rental property and encourage landlords to sell their properties and transfer their investments to more profitable ventures. The imposition of quality standards, which require maintenance costs that cannot easily be met out of increased rents, can have the same consequences. In each case the result is likely to be a reduced supply of housing in this sector; indeed in some countries, notably the United States, some landlords who cannot sell their properties prefer to desert them and leave them derelict rather than maintain them in return for uneconomic rents. Thus policies designed to protect low-income tenants in the private sector often rebound to their disadvantage, because they lead to a reduction in the supply of the needed housing.

Other policies have been introduced to aid both owner-occupiers and landlords to improve their properties, and thus avoid the necessity for urban renewal. These involve the provision of various kinds of improvement grants, both for individual dwellings and for their environments. Such policies have been operating for more than a decade in Britain. To qualify for a grant, however, a dwelling must have an expected life, after improvement, of twenty to thirty years in order to make the investment worthwhile, which eliminates the worst housing. The landlords/owner-occupiers must contribute to the costs; the former will expect to recoup their expenditure in higher rents, which may lead them to introduce higher-income tenants than those removed prior to the work being undertaken, with consequences for the housing conditions of the poorest tenants. (In Britain all local governments are now required to provide for the homeless within their territory, but this is no real solution to the problem of reducing the supply of low-cost private rental accommodation.)

Governments may seek to encourage alternative forms of tenure, particularly for those unable to afford the conventional mortgage market. Housing associations provide one such form. Individuals, charitable organizations, employers or others may form a non-profit-making association whose aim is either to build or to purchase homes which can be rented to members at a price which covers costs but which includes no profit component. Various forms of subsidy may be introduced to encourage such associations, but their small size in many cases makes for great difficulties in policing their activities.

It is in the Third World countries that the problems of housing quality and supply are most acute, and where in many cities the squatter settlements form a major component of the urban fabric. The reaction to such settlements in many countries has for long been that they must be removed, and if necessary their residents sent back to the villages from whence they came. Settlements have been razed, but this usually provides only a temporary 'cure' for the problem because the residents have no rural homes to return to (many were born in the city) and the conventional housing stock cannot provide for them; new settlements soon appeared, sometimes on the razed site. Lack of finance meant that governments could neither provide massive public housing, despite major efforts in many cities, nor stimulate private-sector development. (There are exceptions to this among the more prosperous Third World cities, such as Hong Kong.)

Slowly, therefore, it has been realized that within the constraints of a neo-colonial capitalist economy the squatter settlements present not a problem to the cities of the Third World but a potential cure to the chronic problems of under-employment and poor housing. They offer, it has been argued, an architecture that works, with the occupants investing in their homes as and when they can and in many cases eventually completing a substantial dwelling in 'permanent materials'. The problem is not the housing but its environment, which lacks basic utilities, such as water supply, paved streets, law and order agencies and provision for education. Thus the concept of the planned squatter settlement has emerged. In some cases this involves the provision of 'sites and services' only; instead of squatters moving onto an area, allocating its land among themselves and then allowing the settlement to expand in an *ad hoc* manner, they apply to a government body which has laid out a site, planned a street system and provided certain basic utilities prior to the arrival of the squatters, who proceed to build their homes. The level of service provision varies, and in some more prosperous countries 'core houses' (usually single rooms for cooking and washing) are provided, onto which the settlers add their extensions as and when they can afford the materials. Rents may be charged for the sites, though these are usually low, and loans may be available for the purchase of building materials. Overall, the result is higher environmental standards than are common in the 'unplanned' settlements, a partial 'solution' to the housing problem, and often the creation of a status-ranking of settlements. Furthermore, the public investment in the relatively poor (squatter settlement inhabitants are rarely the poorest in the society) helps to incorporate them into the capitalist system and win their support for its political economy.

Housing classes

In every town and city in capitalist countries, therefore, the market for housing is subdivided into various tenure categories or sub-markets. These are linked in their operations in that, for example, a shortage of properties in one sub-market will probably influence the demand for and price of properties in others. To the consumers, however, there is often little choice with regard to which sub-market they can compete in, and so they are perceived as independent.

The operation of housing markets involves meshing the housing stock, in the various tenurial sub-markets, with the desires of the different social classes. This produces what are termed housing classes, groups within society who occupy different positions within the housing system, some through choice and others through constraint (even coercion in some cases). Eight such housing classes are identified here.

(1) The outright owners of their homes. Most of this group will have achieved their present status after paying off a mortgage, and so are probably at least middle-aged. Some will have been able to make an outright purchase, however; the ability to do this may be related to outside financial considerations, or to a successful property deal, perhaps on a more expensive dwelling in another town. Others may have inherited a home, probably from parents. As owner-occupancy spreads, this last category will probably increase in its size. A consequence may be a reduction in the amount of successful distancing between income groups because inheritance allows people in the middle and even lower strata of the class system to occupy homes in higher-income areas.

(2) The purchasers of homes on mortgages or similar loans. This is probably the largest class in most capitalist societies. They are more likely than the members of the previous class to be segregated by income, because mortgage limits are in most cases closely related to ability to pay.

(3) Renters from private landlords. There are two main types of landlord. The first comprises those who gain their main income from estate management; they may be either individuals or companies and their properties are probably older dwellings in central-city neighbourhoods. Some of the estates may have been custom-built for rental purposes; others may consist of diverse properties purchased as investments. The second type comprises the small landlords, most of whom rent out only one or a few dwellings; many of these will have inherited their properties and use them as an extra source of income rather than as their main financial support.

(4) Renters from public authorities. These can be in one of four types of property, not all of which may be available in any particular country: (i) urban renewal schemes, many of which are dominated by high-rise apartment blocks; (ii) suburban estates of custom-built houses; (iii) older properties, bought by the public authority prior to a redevelopment scheme being put into operation and rented out in the interim; and (iv) specific-purpose dwellings, such as those built for the elderly.

(5) Renters from non-profit-making institutions, such as housing associations.

(6) Those unable to obtain accommodation of their own, who rent parts of dwellings from their occupants. These lodgers are usually either single individuals or kin of the owners/tenants (such as a daughter and son-in-law); their contribution to the household budget may be crucial to its being able to afford that accommodation.

(7) The homeless, who are usually transients, perhaps sleeping rough on some nights and in institutions on others.

(8) The providers of their own homes: the squatters.

Housing market operations and residential segregation

Having identified how housing is provided, and how this provision is meshed with the class system, the next task is to investigate how these processes operate to produce and maintain the desired residential separation of the classes. The following sections look at each of the major tenure categories in turn.

Ownership, mortgages and class segregation

Because of its longevity, as well as its size, a home is an expensive commodity, and for most people its purchase is dependent on being able to borrow the majority if not all of the price. Money-lending institutions are central to the operation of this tenure category, therefore. These are not organized as philanthropic bodies, despite their major aim of advancing the cause of home ownership. They are capitalist concerns, and it is their search for profits which determines their policies and hence the social morphology of urban areas.

The main financial institutions operating in the housing market have been created specifically for the task of providing mortgages, and this is their only function. Their role is like that of the merchant: they bulk the needed commodity, in this case money. Those with money to invest are attracted to deposit it with the institution, to finance loans to would-be home-buyers. The aim of the institution is not to make a profit for its owners in the normal sense, for it is the depositors and not the managers who are in fact the owners. But in order to attract deposits the rate of interest offered must compete with the rates available for other forms of investment; thus the rates are, in effect, the profits. Other advantages may be offered to depositors as well, such as interest paid on all deposits, however short term, which is not the case with most banks, and the tax relief on interest received which British building societies have obtained for their depositors. The building societies in Britain and the savings and loans associations in the United States are archetypal of these financial institutions which are exclusively concerned with the housing market. The former began as self-help organizations for neighbourhood groups, and there are still many small ones serving local areas only; some are now large, nationwide operations, however, handling millions of pounds annually and advancing several thousand mortgages. The term 'building society' is used in the rest of this discussion to apply to all such institutions (and private individuals who lend their own money, often through solicitors, on similar terms).

A major characteristic of building societies is that they 'borrow short and lend long'. Most of their deposits are on instant recall, but their loans are for long periods, many for twenty-five years or more. Thus they cannot afford even the slightest hint of unsteadiness in their operations, for if a large number of depositors suddenly demanded their money the societies could not repay them, unless they could borrow the sum elsewhere. They must give every indication of security, therefore, radiating sound management and sensible lending policies.

Most building society mortgages use the property concerned as the collateral. The interest rate charged must be sufficient to cover operating costs and pay an attractive interest rate to depositors; the society must be sure that the interest payments on its loans will be made regularly. This requires absolute confidence in both the clients and the properties that they buy. Confidence in the client is ensured, as far as is possible, by lending only to those who will be able to meet the obligations of the loan, thus excluding the low-paid and the under-employed who receive insufficient income. Further, the society will set an upper threshold on how much it will lend to any one person, again based on his capacity to pay; this is usually some ratio of his regular basic income (excluding bonuses, overtime payments and even, in some cases, wife's income). Thus certain classes are excluded from being potential borrowers from the society, and are denied access to certain housing classes. (It is interesting to speculate to what extent those excluded use the societies as banks for the deposit of their small savings.) Above those classes, the amount of mortgage available is closely allied to income, although the society will often take into account the income prospects of, for example, a professional person whose earnings are on an incremental scale. Thus the value of property one can buy is closely circumscribed by income, which limits where an individual can live.

Confidence in the property on which the loan is made is ensured by limiting those for which money will be advanced. (The role of the property as collateral for the loan is important in this context, since if the society has to foreclose on the loan, the property then reverts to it, and must be sold at an acceptable price in order to recover the investment.) Thus old and depreciated dwellings are rarely considered to be good investments, since they are unlikely to hold their value over time. The building societies focus their loans on properties which are likely to maintain, and perhaps increase, their relative value. Their main interest is in new homes, therefore, including in this category those which are less than about twenty-five years old. They may also be prepared to lend on properties in areas of 'social improvement', where it is clear that for some reason property values are on the increase, but they will shun areas of depreciating properties and many negative externalities. Indeed, the practice of 'red-lining', by which certain areas of a city are designated to receive no loans, is widely observed, implicitly if not explicitly.

The building societies and comparable institutions are thus organized to finance the home-buying of those classes within society whose members are characterized by secure and sizeable incomes and a desire to live in the sorts of areas where property values are unlikely to decline. Their investment policy is thus a prudent one. In this they are supported by a variety of other institutions. Insurance companies, for example, accept some of the risk by offering cover on the lives of the mortgagees, cover which will meet the full cost of the mortgage if the borrower dies. They also invest money in the building societies, as do pension funds and similar bodies; clearly all of these are concerned that the societies should operate sound lending policies so that the returns to the investors are secure.

Estate agents are also associated with the building societies as major gatekeepers controlling the house-purchase system. (Some of them are very closely associated, in that they act as agents for the building societies.) At least a quarter of all home purchases are arranged through estate agents, whose payment consists of a percentage of the selling price. The agents are also concerned to maintain property values in areas, therefore, and they attempt to achieve this by channelling clients to what they consider to be the 'right' areas for people of such income and status. They are most careful to avoid introducing certain negative externalities to neighbourhoods (except in special circumstances which involve them in speculative dealings), and in the United States realtors have for long operated a 'code of practice' designed to maintain the characteristics and qualities of neighbourhoods.

All of these policies of the major institutions involved in the mortgage market combine to discriminate in favour of the more affluent members of urban society, and hence the 'better' areas of the town. They are major forces acting to produce and, where necessary, enhance the desired patterns of residential segregation. But what of the people to whom they will not grant mortgages, and of the areas they will not invest in? Some of them may be provided with public sector housing and others may be able to rent from a private landlord. Others find that they must buy, however, and to raise the necessary money go to a range of money-lenders, including private individuals and, in some countries, local governments. To provide such mortgages to people who have been classified by the building societies as relatively poor risks, and perhaps also on dwellings which have been similarly categorized, these money-lenders have to raise their funds at relatively high rates of interest, competing with the building societies, who have certain advantages over them with regard to tax concessions. As a consequence, the mortgages from money-lenders are usually for relatively short periods and at high rates of interest so that the households who are dependent on them, and are already constrained by their low incomes to the less desirable properties and areas, have to pay relatively more for their homes than do those who borrow from the building societies. In this way, the class structure is maintained not only through residential segregation but also by 'keeping the poor relatively poor'; many of the latter have to sublet parts of their homes in order to meet their mortgage commitments, and so live in cramped, often unhealthy conditions.

One final group involved in the housing market who are associated with the building societies and other gatekeepers are the developers, the producers of new housing. As capitalist firms, these are interested in the profits they can make, and so they tailor their offerings to the requirements of the major lending institutions; indeed, in several countries developers make arrangements with building societies, prior to beginning any construction, that the societies will grant mortgages on the proposed homes, and in some cases they finance the development with loans from the societies. Thus the social categorization of new residential areas is virtually assured; developers are rarely interested in building for the lower-income classes, and they focus their activities on areas close to the existing high-status neighbourhoods.

All of these operations both preserve residential segregation and stimulate the circulation of capital; the latter is part of the maintenance of the affluence of those who invest in the construction and associated industries. In addition, other patterns of spatial separation are encouraged. At the beginning of this chapter it was reported that geographical analyses have indicated some separation of the age groups within urban areas. The building societies prefer to grant loans to the relatively young, who want new homes for their growing families. The older you are, the harder it is to get a mortgage, so that the new suburban developments are dominated by particular age groups (mainly 0–10 and 25–40). Despite very high levels of mobility, a neighbourhood's population tends to age with the homes, so that the older the area the older its population. As a result there is a clear separation of the generations within cities, with a close correlation between age and proximity to the city centre.

Occasionally, the character of an area changes, sometimes quite markedly, over a relatively short period. In class terms this may indicate a decline in its status, as one group succeeds another; alternatively, though less frequently, the change may involve an increase in the area's status, with the incoming population being of a higher class than that being replaced. The various institutions discussed in this section are frequently involved in such change, and may indeed be its catalyst. The decline in status, for example, may reflect the interpretation of estate agents that an area is beginning to depreciate, and that they should accelerate this process. The classic case of this involves black movement into formerly white areas of American cities. Once a few homes in a neighbourhood have been sold to blacks, the white residents may perceive the build-up of a major negative externality and prepare to leave. Estate agents may help to persuade them of the need to move; the panic selling thus generated may allow the agents to make considerable profits, by purchasing homes cheaply from the whites who are keen to depart and selling them at much higher prices to the incoming blacks.

Examples of the other sort of change include the process which is now widely termed 'gentrification'. Here an area occupied by low-income households is 'invaded' by those of higher-class position, who invest in an upgrading of the properties, which are usually relatively old and close to the city centre. The stimulus for such developments is usually the desire of members of certain professional and other classes to live close to the city centre, and this desire is advanced, often promoted, by local estate agents and building societies in the area to be gentrified. These see that large profits are to be made in buying relatively inexpensive low-class residences and selling them to higher-income households at a premium, often after repairs and other works have been carried out. (In Britain part of the costs of such improvements may be met by grants designed to aid the regeneration of inner areas, but not to stimulate profiteering and the takeover of neighbourhoods by the affluent. In many cases the areas being taken over were once the homes of the affluent, but have since filtered down the class system as a consequence of suburbanization.) The process involves the creation of a new set of positive externalities in the area, which define a new social ambience and attract households who do not favour suburban living; Chelsea and Islington in London, Georgetown in Washington and Paddington in Sydney are illustrations of this.

Other institutionally stimulated, if not generated, changes include those aimed at new lifestyle groups within urban society. The development of semi-communal lifestyles – for both the relatively young and the relatively old – has been catered for by apartment complexes and condominia, in which co-ownership arrangements are made for the communal parts of the property. Such developments are characteristic of North America. In Australia and New Zealand the trends in the last decades to earlier marriage and fewer children have meant that there are many parents in their late forties whose children have left home and who do not wish to spend the next twenty-five to thirty years of their married life in a three- or four-bedroom house set in a quarter of an acre of garden. And so developers are finding that there is great demand for smaller, usually two-bedroom ownership units (OYOs) among this group within the urban population.

Allocation of public housing

In most capitalist societies public housing is occupied by those households which do not qualify for mortgages and home ownership and for whom the private rental sector offers few opportunities: families with young children are often discriminated against by private landlords, for example. Thus although in theory public housing may be available to all, an excess of demand over supply means that the managers (another group of gatekeepers) must determine criteria which discriminate between different groups of applicants. These managers may operate on quasi-capitalist criteria. Much of the early public housing in the United States, for example, had to meet all of its operating and maintenance costs out of rental income (the housing was built with federal grants); in such circumstances, to protect their own solvency housing authorities tended to select tenants who could afford the unsubsidized rents, thus leaving the poorest classes to fend for themselves in the private sector. Since 1969 federal funds have been available to subsidize rents, and tenants are now not required to spend more than one quarter of their income in this way; this has opened up public housing for the poorer classes.

British local authority housing provision has always involved a strong welfare component, substantiated by legislation passed in the mid-1970s which makes local governments ultimately responsible for housing all of their residents. Because demand for their homes exceeds supply, all authorities have had to evolve rules for allocating the available dwellings, for deciding who to allow onto the waiting list, and for placing applicants on that list. Priority must be given to households whose current dwelling is to be removed under some form of slum clearance programme, as authorities are required to offer alternative accommodation to these. Otherwise, the rules vary somewhat from authority to authority, but in general the major criteria employed to allocate points to applicants are: young families, especially those whose children have to share bedrooms with their parents; households in poor living conditions; households who have to share certain facilities, particularly cooking and toilet; people with medical support for living in a certain type of accommodation; and long-term residents in the authority's territory (many will not house those with less than five years' residence, which militates against immigrant groups). Owner-occupiers, tenants of satisfactory privately rented accommodation, and small, especially older, families are the least likely to be offered accommodation, or even to be allowed onto the waiting list, so that the characteristics of council housing tenants are not necessarily those of the poorer classes as a whole.

Within the public housing sector, and especially in authorities with a large housing stock, there is often a considerable range of accommodation, in size, type (flat or house), age and condition; the last two are strongly correlated, with the older properties in the poorest condition. In managing this stock, authorities define policies akin to those of private capitalists, because they must raise sufficient rent income to cover a large part of the costs of maintaining the homes and paying for the loans which financed their construction. Rents are thus usually higher for the newer properties, which are then offered to the more prosperous families. Immediately this creates class segregation, between estates and even between different parts of the same estate, and the higher-income residents of the 'better' estates will seek to protect such differentials and maintain their own status within their particular housing sub-market.

Some of the households which authorities are obliged to offer homes to they classify as 'problem families', usually with regard to either recalcitrance over rent payments or poor upkeep of their homes. (The latter creates negative externalities, lowering the perceived quality of areas and encouraging the 'better' tenants to seek transfer to other, 'better' estates.) Before offering a dwelling to a family, therefore, an authority will assess the degree to which it is likely to be a good tenant; this is usually done by observing the cleanliness and upkeep of the current home. Those perceived to be potential problem families will only be offered relatively low-quality dwellings on the older estates, thus creating class and other differentials between areas within the public housing sector. Tenants of other council properties who turn out to be 'deviants' may be reallocated to similar areas, which rapidly gain reputations as dumping grounds for 'problem families'. (Many of these families are large, and the relatively small size of the houses – there are few with more than three bedrooms – means that their children are constrained to spend much of their leisure time on the streets, with consequences in their behaviour for area reputations.) In attempting to operate their stock on 'good management' lines, therefore, both the employees of the local housing authorities and the politicians who oversee their activities create patterns of intra-class segregation. There is a clear status and desirability ranking of public housing estates in most towns and cities.

In state capitalist societies the majority of the housing stock (the entirety in a few, at least in the urban areas) is in the public sector. Its allocation involves managerial decision-making, as with public housing in capitalist societies, but the ideology guiding the state capitalist managers is a different one: equality of provision relative to need.

The current state capitalist societies inherited a considerable housing stock from the mercantile and capitalist societies that they replaced. This stock contained major differentials, both within and between the various housing classes, and was spatially segregated. Such differentials were removed in some, as far as was possible without demolishing the needed dwellings: in China, for example, the former middle-class homes built around courtyards were subdivided and allocated to several households, who shared many of the dwelling's facilities. But in others private ownership was retained, although landlords in the private rental sector were dispossessed and their dwellings transferred to the state. New housing – much of it in apartment blocks – was built by the state only, but concentration of investment on industrial expansion has meant a considerable housing shortage in most state capitalist cities. In Hungary the professional and skilled workers associated with the new industries were placed at the top of waiting lists for new homes. Others were less fortunate and were constrained to the older, generally poorer rental property. Many still live in villages, and commute long distances to the urban factories, perhaps sleeping for several nights a week in a workers' barracks.

The rationale for such inequalities in housing provision under state socialism in Hungary has been provided by Ivan Szelenyi (in his book *Urban Inequalities under State Socialism*). There was a post-1945 shortage of housing in Hungary, as of other consumer goods. Wages were set to preclude a sizeable private market in housing, and so the vast majority of Hungarians had to rely on state provision. The supply of new housing could not meet the demand immediately, and bureaucratic procedures were evolved to allocate that available. The most worthy applicants were judged to be the key workers (who also obtained the highest incomes). As Szelenyi puts it 'How could it be otherwise. How could the state say to its rising managers and bureaucrats, "If you get promoted you will *reduce* your housing chances?" ' Thus in Hungary over the period to 1968, it was the high bureaucrats, the intellectuals, technicians and clerical workers whose housing conditions improved relative to national norms (they had bigger homes, occupied at lower densities, and with better facilities) whereas the manual and service workers experienced relative declines in housing quality. Despite the proclaimed policy of providing housing relative to need, therefore, the managerial operations ensured inequality.

Other pressures have led to considerable inequality in housing provision and a consequent segregation of classes in state capitalist cities, as has been illustrated by Bater for the Soviet Union. There a national minimum housing standard of nine square meters of living space per person has been set, but only in a few cities does even the average reach this target. (Most of those cities are in the west, suggesting the operation of policies which discriminate against the ethnic minorities of the east.) The state sector comprises some 70 per cent of all housing. It is divided into two main types: that provided by the city soviets and that provided by individual enterprises for their own workers. Many of the latter need to attract labour so that the apartments which they build (some 60 per cent of the state total) are frequently superior to those provided by the local governments. Thus there are housing differentials and class (or strata) segregation according to employment, with advantages to the employees of certain factories, plus many of the higher bureaucrats. The other 30 per cent of the housing stock is in the private sector, consisting mainly of owner-occupied homes, which can be traded. An increasing proportion (15 per cent of all current construction in Moscow) is in cooperatively owned apartments. The owner-occupiers must provide a 60 per cent down-payment, and the cost of the 40 per cent state mortgage is much greater than the very small weekly rent usual for a state apartment. Thus only the relatively affluent – mainly in the upper white-collar strata – can afford to segregate themselves in this way, in many cases in cooperatives organized by and for particular occupational groups. The result is not different in general form, though there are many variations in detail, from the housing markets of capitalist cities.

Housing markets and housing choice

The discussion of housing markets so far may be interpreted as showing that households are pawns in the hands of the political–economic system. Their occupation places them in a certain social class and income group, and this determines their access to the various housing submarkets, and thus the parts of the city in which they can live. As a broad generalization this is accurate, but it must be tempered. Within several of the housing classes households have certain freedom of choice, in part because of the size of the market in which they can compete and in part because of the flexibility they have in allocating their income between housing and other items of expenditure. Although the housing class boundaries may be relatively impervious, therefore, within them there may be considerable freedom of choice; in general, the greater the income and power of a class, the greater the choice for its members.

The exercise of choice in the owner-occupier and mortgage-holding housing classes, and to a lesser extent among tenants of private landlords, is frequently researched through investigations of intra-urban migrations. Both the decision-making involved and the pattern resulting from a large number of moves have been popular topics for urban geographers. The majority of people who move within a city, it appears, do so because of changed household requirements with regard to living space: most of them want a bigger dwelling. They may, like the squatters, meet this requirement by building extra rooms onto their existing homes, and those who want a smaller dwelling may divide up their home into several apartments. But most decide to search for another. They define the sort of dwelling that they want, and also the area in which it must be located, and then search the available vacant stock for one which meets their criteria, suits their resources and is acceptable to their financial sponsors. Many searches are unproductive, for a variety of reasons, and lead the would-be movers to decide to adjust their present homes to meet their demands. But many find what they are looking for, usually within a short distance of their current home, so that the pattern of intra-urban migration is dominated by short-distance moves. This pattern is in part 'explained' by a desire among movers to stay in the same general neighbourhood, where they have well-developed social networks; in part it reflects the problems of searching over wide areas, including districts of the city about which they know very little and whose potential positive and negative externalities they are unable to assess; and in part it indicates the constraints set by the spatial segregation of the various housing sub-markets. Thus, as with the choice of friends discussed in the previous chapter, it is difficult to unravel the pattern of moves into a causal sequence: basically the pattern of residential segregation of the housing classes limits the available opportunities for a household with a certain income to particular areas of the city only, but within those areas most movers choose homes in the districts that they know and have contacts in, close to where they previously lived.

Some people move for reasons other than space requirements. It may be that a rise in their real incomes allows them to aspire to higher housing standards; alternatively, either the operation of the filtering process may make better-quality homes available or negative externalities in the neighbourhood may convince them of the need to move in order to maintain their position in the status system. Such moves may well involve a change of neighbourhood, as might those for people who are being forced down the class system by changes in their absolute or relative economic status. But both groups are likely to prefer their 'known' portion of the urban area, unless the change in status is transferring them to a very different sub-market. And finally, a few may want a major change of environment – another part of the city, perhaps, or a different type of dwelling – for which they are prepared to move a long distance.

Turnover is in general much greater in the private rental housing class compared to the owner-occupier, in part because it is easier to give notice to one landlord and to rent a property from another than it is to make the legal and financial arrangements for buying one home while selling another. Moves in the sub-markets just discussed are usually the result of much careful consideration, therefore, whereas in a sector characterized by a large proportion of young and relatively footloose households, and also of low-income families who lack security of tenure, the decision to move may be a less weighty one. (It may, of course, be an imposed one, if the decision is the landlord's rather than the tenant's.) Because much of the housing involved is concentrated in the older housing of the inner city, the result is a complex pattern of short-distance moves within this area, representing the frequent readjustments between housing requirements and provision. Often the search for an alternative home is made under severe time constraints, which prohibit any wide and careful evaluation procedure and encourage later movement consequent on a poor, enforced choice.

Finally, the public housing sector tends to have low turnover rates, although this varies somewhat between societies depending on the flexibility of the housing managers and the ratio of supply to demand in the sector; intra-urban migration is generally very slight in state capitalist cities, for example. Many tenants wish to move: some need extra space, especially bedrooms; some aspire to live on 'better' estates or in different types of dwelling – a house rather than a flat is usual; and some need to be in other parts of the town, for access to workplace or to a dependent relative, for example. Authorities differ in their attitudes to such demands, and in the degree to which they encourage and handle applications for transfers. Most give priority to those who have medical support for their request, and many are sympathetic to those with severe overcrowding problems. Of the others, it is usually the 'good' tenants, who pay their rents regularly and maintain their homes in good condition, who have the best chance of being provided with the desired transfer. But most authorities have long waiting lists, and households on these must take precedence. In Britain, for example, those who are being offered council homes as part of slum clearance programmes are able to demand a high-quality, new home as their price for moving out of the condemned property, and those with medical support are able to push their claims. The 'average' tenant asking to move to a suburban house, perhaps from a tower block, may well find that these other groups within the housing class are more powerful and able to move to the 'better' properties, so that his or her chance of achieving the desired transfer is minimal.

Many intra-urban moves represent the operation of choice within the housing classes, therefore; the main exceptions are those evicted in the private rental sector, the mortgage defaulters in the ownership sub-market and the compulsorily moved in the public housing sector. There is also some movement between housing classes. Together these moves involve between 10 and 20 per cent of the population of each urban area annually, in cities of late capitalist countries at least. In mapped form, the pattern of moves reflects the existing pattern of spatial segregation of classes and the constraints on movement within and between these classes. As a result, the large volume of movement leads to few changes in the characteristics of areas, at least in the short term: those moving into an area are very similar to those leaving it, a generalization which also holds for those moving between cities, usually as part of their career cycle.

The spatial pattern of residential areas

Although the whole of the discussion in this chapter has been about social areas, the differences between them and how these are created, nothing has been said about where these different area types are located within the city's morphology. Much geographical research effort is concerned with the identification of regularities in spatial patterns, and the residential mosaic of urban areas has been a major topic within such a paradigm. Thus this final section of the chapter looks at the regularities in the spatial organization of residential segregation.

Studies of such spatial organization are based on maps, and cartographic presentation of where people of different individual and household characteristics live is far from novel. The massive work of Charles Booth on poverty in London in the 1890s included maps showing the condition of every dwelling, and he generalized from these in his conclusion in noting that 'residential London tends to be arranged by class in rings with the most uniform poverty at the centre'. More recently academic social scientists, including urban geographers, have undertaken a great deal of map analysis, mainly of data collected in censuses; the bases for their efforts have been provided by the pioneering speculations of a few sociologists and economists.

Zones and sectors: The sociological arguments

Credit for the first attempt to formulate a general statement about the residential patterns within cities is usually given to the Chicago sociologist Ernest Burgess. During the 1920s he published a series of papers based on his detailed field and statistical investigations of Chicago, in which he suggested that the typical North American city could be divided into a series of zones (much as Booth had suggested for London). His model, which is clearly founded on Chicago (Figure 7.1), presented a synopsis of the whole built-up area, but most of his research was concerned with the concentration of various symptoms of social disorganization in the central zones; he wrote very little about the outer zones. It is perhaps for this reason that, as Figure 7.1 shows, his various papers were somewhat ambiguous with regard to the zonal terminology, but study of his writings indicates that he was clearly suggesting a straightforward relationship between distance from the city centre and the class composition of residential areas; the further from the centre, the higher the class there. Although so obviously based on Chicago, Burgess in-

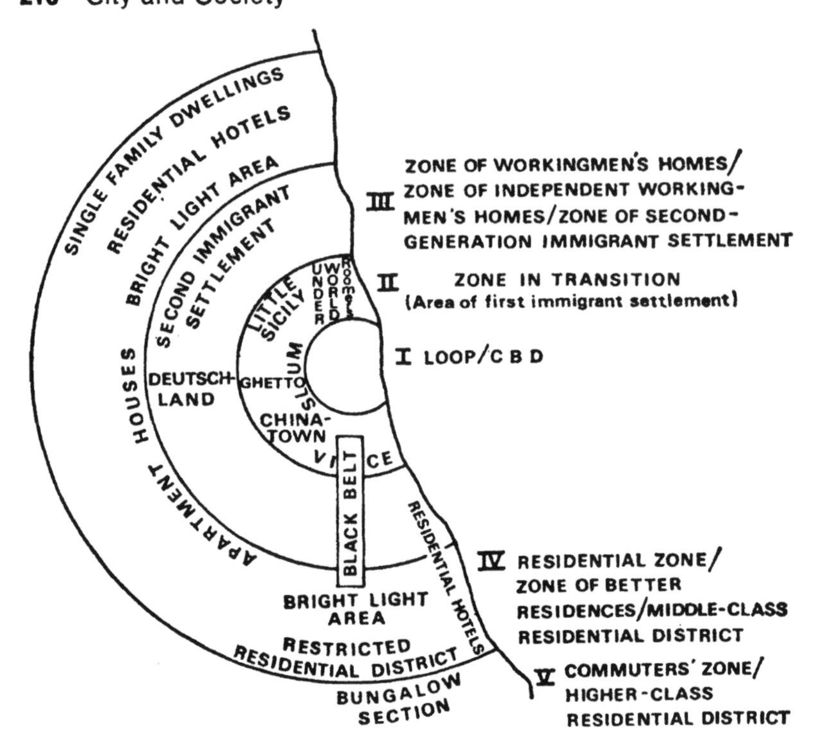

Figure 7.1 The Burgess model of the spatial organization of residential areas in North American cities, based on three of his papers. The several names of the zones indicate the variations which he introduced in his descriptions.

tended his model (or idealized diagram) to apply to all cities in the United States.

Burgess's zonal schema was modified in the late 1930s by a land economist, Homer Hoyt (also with academic roots in Chicago), who derived a general statement about urban spatial organization from a detailed study of residential patterns in 204 United States cities; the work was commissioned by the Federal Housing Administration to assist its decisions on mortgage insurance. Hoyt did not deny the existence of a zonal organization, but argued that his evidence indicated that a sectoral pattern was the dominant feature in the set of cities for which he had maps. (Burgess's map – Figure 7.1 – hints at the existence of a sectoral pattern, notably in the 'black belt'.) The different classes tend to be separated out into different sectors of the city, with the highest classes pre-empting the most attractive environments, usually those with either or both of relatively hilly topography and a pleasant waterfront. Within each sector, there could well be a zonal pattern too, as the result of the operation of the filtering process (p. 195). Unlike Burgess's model, therefore, Hoyt's was less precise in that the relative location of each sector could not be predicted for the 'average city'. Thus Hoyt did not present a cartographic representation of his model to parallel Burgess's; that portrayed in Figure 7.2 is the most common representation of his ideas, one that Hoyt has never repudiated.

Both Burgess and Hoyt aimed to describe not only the patterns within cities at one time but also the processes which produced changes and yet maintained the same morphology outlined in Figures 7.1 and 7.2. Burgess's work was set in the tradition of the so-called Chicago school of human ecologists, whose ideas on social organization were strongly conditioned by the biological analogy which they drew: this was a form of social Darwinism based on ideas of the survival of the fittest in the competition for space, and Burgess and his colleagues argued that sociologists could learn much from studies of the plant and animal communities. Thus for Burgess the dominant process of change within the city was one of invasion and succession, involving the movement of migrants to the city. Most immigrants are in the lowest classes, and so are constrained to the older rental housing of zone II in his diagram. As their numbers there

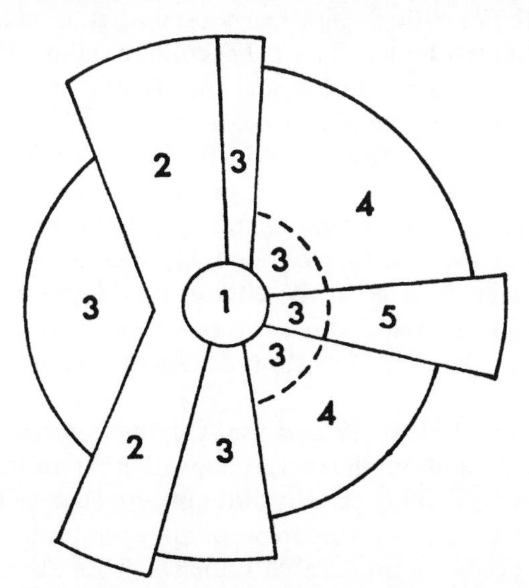

Figure 7.2 The commonest representation of the Hoyt sectoral model of the spatial organization of residential areas in North American cities. Key: 1 – CBD; 2 – wholesale and light manufacturing; 3 – low-class residential; 4 – middle-class residential; 5 – high-class residential.

increase so the other residents, and also those of adjacent neighbourhoods, feel that the quality of their environment is being threatened and so they seek to move into housing in the next zone. (Such movement blurs the inter-zonal boundaries, and in some of his papers Burgess talked of continuous gradients rather than discrete zones.) Thus not only do the immigrants introduce negative externalities to zone II but those that they displace have a similar effect on zone III. In their turn, the residents of zone III seek to move into zone IV, so that the initiation of the invasion process in the inner city stimulates a rippling effect which reaches out to the edge of the city where the highest classes are constrained to build new suburbs beyond the current built-up area in order to protect their residential exclusiveness. Thus the invasion of one neighbourhood by an immigrant group, and its eventual succession as it comes to dominate the local population, generates a large number of consequent invasions which disturb the equilibrium of the residential pattern.

Hoyt accepted the logic and existence of invasion and succession sequences, but argued that other processes also operate to produce change in residential area characteristics. Thus high-class residents in the outer zone of a sector may decide to move out to a new suburb being built further from the centre, not because of the introduction of some negative externality to their existing neighbourhood but rather because of a belief that they needed new homes. This belief could be stimulated by a variety of factors, including the technological and stylistic obsolescence of their current dwellings; the notion of obsolescence may be introduced by the capitalist firms in the construction industry whose continued profitability depends on them being able to sell expensive new homes to the wealthy classes. In moving for such reasons, the upper classes initiated filtering processes within their sectors of the city, so that whereas Burgess's rippling effect was initiated at the city centre, Hoyt's originated in the edge of the built-up area. Both are represented by the single mapped pattern of moves, dominated by short distances and an orientation away from the centre.

Both Burgess and Hoyt realized that they were presenting very high-level generalizations about the morphology of urban residential patterns, and that they were making certain assumptions, such as that the central business district (CBD) was the focus of all commuting and other trips. Nevertheless, despite considerable criticism from some quarters, the models were widely accepted and adopted and formed the basis for the description of patterns in many cities, initially in North America only but, after the Second World War and adoption of the models by geographers, in several other countries too. The bulk of the mapped evidence supported the hypotheses regarding both zones and sectors, and patterns of invasion and succession could be identified in studies of the waves of immigrants to American cities. The operation of the filtering process was much less apparent, however. Many of the vacated homes in suburbia were occupied not by former residents of homes closer to the city centre but by either high-class immigrants from other cities or newly formed, relatively affluent households. Chains of moves were initiated by suburban construction, but these rarely reached back into the inner city and few members of the poorest classes benefited as a result of the construction of new homes for the rich.

During the last two decades the wealth of census data for small districts within cities has presented the means for widespread testing of the Burgess and Hoyt models. Most of the findings for North American cities have shown that, with respect to the three main types of distancing identified in factorial ecologies (p. 188):

(1) The map of social class distributions is largely sectoral in form, with zones within the sectors. In most cities, however, the zonal arrangement of the highest-status sectors does not conform with the models, for the high-class residential areas tend to be neither on the edge of the built-up area nor adjacent to the CBD. The reason for this is taken up in a later section.

(2) The map of household types and age structures is dominantly zonal, with young families on the city edge, older families in the intermediate zones, and apartment- and flat-dwellers, most of them not members of nuclear families, concentrated in the inner zones. (Some analysts claim that this pattern was suggested in Burgess's writings; he called the fifth zone in his model the 'domain of the matricentric family', for example.)

(3) The map of minority groups shows that each lives in a zonal-cum-sectoral concentration, forming a cluster which is usually zonally organized with regard to socio-economic status and household types.

Similar findings have been reported for other late capitalist countries, notably those of the English-speaking world. Each has its own peculiar features, such as the large public housing sector in British cities, but the general arrangement is very much that proposed by the American models.

Zones and gradients: Economic arguments

Alongside the sociological description of typical urban patterns presented in the previous section is one produced by economists, using a different modelling procedure. The origins of this second attempt to explain urban residential patterns lie in work by von Thunen on land use in the 1820s and by Hurd on urban land values in the early 1900s; detailed expositions were only developed in the 1960s, however.

Assume a city in which each household contains one member of the workforce, whose place of employment is either in or close to the CBD. Each household's budget can then be divided into three categories:

L = expenditure on land and housing;

C = expenditure on commuting; and

O = expenditure on all other goods (including savings).

Thus

Total expenditure = L+C+O.

If for the moment it is assumed that O is the same for all households, then the remaining sum (L+C) can be allocated in a preferred division between property (L) and commuting (C). Assuming that (L+C) is constant for all, then since commuting costs increase with greater distance from the city centre, the location of all employment, then the amount available for spending on land and housing will decline with distance from the CBD (Figure 7.3A). This produces what is known as a bid-rent curve (Figure 7.3B),which shows the maximum price people are prepared to pay for property with increasing distance from the city centre. Because commuting costs do not increase as a linear function of distance travelled (the bus fare for a four-mile journey is less than twice that for a two-mile journey) and the amount of land available increases with the square of the distance from the city centre, the shape of the bid-rent curve is usually that of an inverted J, as indicated in Figure 7.3C.

Where on this bid-rent curve will each household choose to locate its dwelling? Those who want to be close to work and to keep commuting costs down will select a relatively central location, whereas those not so concerned with commuting costs (probably including the time element) will substitute the higher travel costs of distant suburbs for the high property prices closer in. It has been argued that in general it is the poorest who will occupy the more central locations, for the following reasons.

(1) All people would prefer as much space for their home and garden as possible.

(2) Land is cheapest at the edge of the built-up area, so that for any fixed sum of money the size of property that can be bought increases with distance from the city centre.

(3) If the category O in the earlier equation is subdivided into

N = expenditure on necessities; and

E = expenditure on 'extras'; so that

O = (N+E)

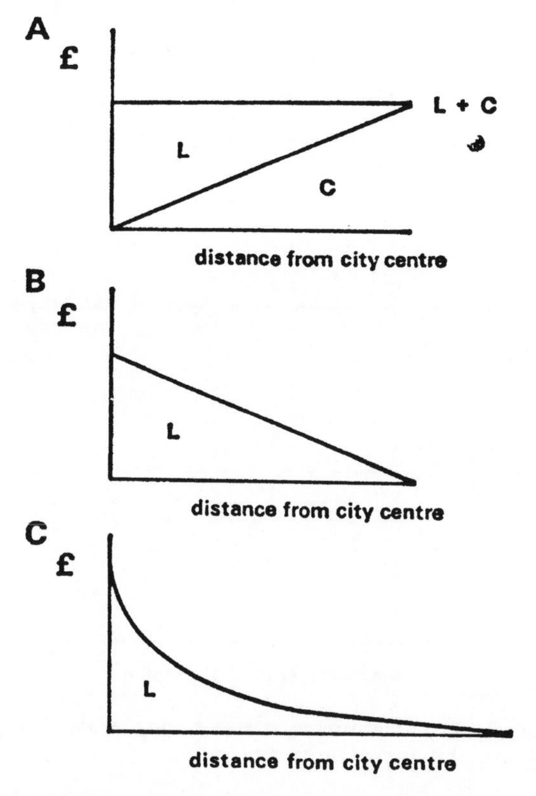

Figure 7.3 The derivation of bid-rent curves. L=expenditure on land and housing; C=expenditure on commuting.

then the richer classes will have a greater volume of E available, part of which can, if desired, be allocated to the purchase of space. As they have less available in the category E, the poor cannot outbid the rich for the larger properties and so are confined to the inner areas.

(4) Commuting costs do not vary markedly between the various classes, but for the richer classes they are relatively less important in their total budget. As a result the richer are less sensitive to increases in commuting costs and better able to live considerable distances from their workplaces. Figure 7.4 illustrates this for two income groups, one earning £X and the other £Y. The shaded area indicates the effect of a 25 per cent increase in commuting costs. The further out a household lives, the more it has to absorb from this increase. For the low-income groups this means either reallocating some of their expenditure on L to C (which means accepting lower housing standards and higher densities) or cutting some

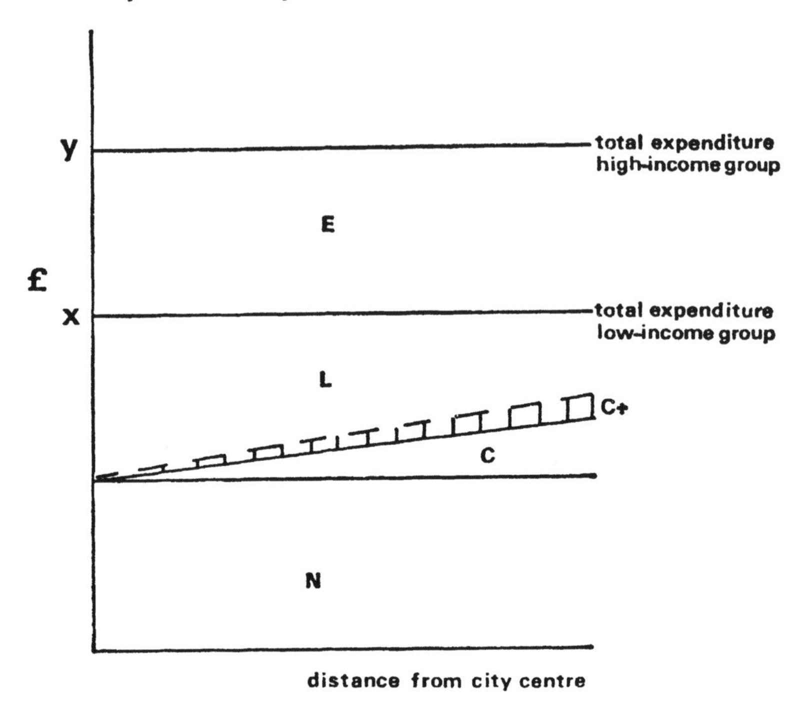

Figure 7.4 The effect of an increase in commuting costs (C+), in real terms, on the expenditure of two groups, **x** and **y**.

of their expenditure on N: they have no expenditure on 'extras' (E) which can be reallocated to C. For the higher-income group (Y), on the other hand, the extra cost (C+) is easily absorbed in E without having any substantial impact on other living standards.

Two deductions can be drawn from these arguments: that population densities should be greatest at the city centre (excluding non-residential areas) and should decline away from the city centre in an exponential form (as in Figure 7.3C); and that average household incomes should increase with distance from the city centre. Both have been tested and confirmed many times, for cities in many capitalist countries, with modifications to relax the assumption of all employment in the city centre. Thus the rich, it is argued, are price-oriented, getting the space that they want where it is cheapest; the poor, on the other hand, are location-oriented, unable to afford the space on the edge of the city and so living in the inner-city areas where they pay low commuting costs.

This economic argument has been used to bolster the Burgess and Hoyt models. Its validity depends on the correctness of its assumptions, however, for it ignores consideration of how housing markets operate and how the rich use their power to impose their preferences on the urban mosaic. In fact, the high-income classes live on the edge of the city – if indeed they do – not only because of a demand for space (which is far from infinite) but also because of a desire to avoid the negative externalities of inner-city locations. If the higher-income groups had other desires with regard to space and perceived negative externalities, and wished to live elsewhere (as suggested by the gentrification trends) then their greater purchasing power would allow them to oust the lower-income groups, who are always constrained to living where the more affluent prefer not to. Figure 7.5 illustrates this for the two hypothetical income groups of Figure 7.4. In the first case (Figure 7.5A) the higher-income group (Y) want space and suburban residence; of their £E, therefore, they spend £S on space in the outer zone and the lower-income group (X), lacking the £S, live in the inner zone. In the second case (Figure 7.5B), the higher-income group prefer to live close to the city centre, and they will pay £S for residences there. The total amount they spend on property in the inner city – £(L+S) – exceeds that available to group X, who have three possible courses of action open to them:

(1) they can continue to compete for properties in the inner city, but because they can only afford £L this means that they must accept very low housing standards and very high housing densities;

(2) they can reallocate part of £N to L – up to a maximum of £N* – accepting lower other living standards in order to afford housing in the locations they prefer (they may be the only locations they are allowed access to); and

(3) they can accept defeat, and live in the outer zones, in which high commuting costs are involved and the implications of Figure 7.4 hold. Whichever is chosen, the result is lower living standards for the lower-income group, as a consequence of an application of the economic power (probably bolstered by political power) of their affluent peers.

The patterns observed by Burgess and Hoyt are not 'naturally' determined by economic and social forces, therefore. Rather they reflect the preferences of those with economic power in the housing market, who are able to manipulate the housing market in order to facilitate exercise of those preferences. If their preferences are not for space and they are not prepared to accept commuting costs and time, then they can impose those on their weaker brethren. To understand residential patterns involves an understanding of preferences and their implementation in the capitalist system, which leads to the presentation of an evolutionary model.

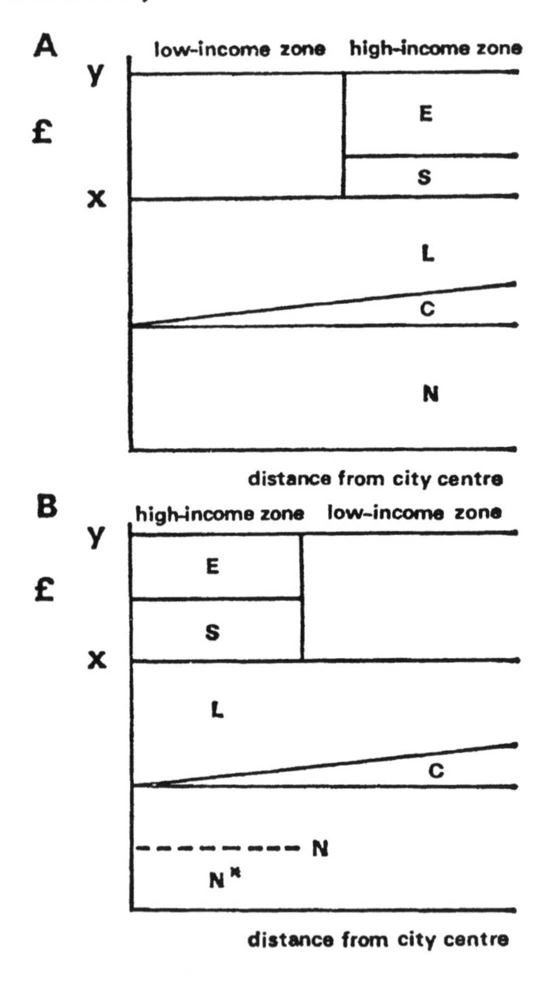

Figure 7.5 The effect of changed preferences by the higher-income group (y) on residential location patterns.

An evolutionary approach

Analyses of the spatial organization of urban social areas, both for different times and for different places, suggests that the various stages of capitalist evolution are characterized by particular arrangements. This has led to the idea of an evolutionary sequence comprising an initial stage with the higher-income classes living close to the city centre, a second stage which is the reverse of that pattern (and thus conforming with the Burgess and Hoyt models), and a final stage in which the low- and high-income groups are outflanked by the development of middle-class suburbia. In detail, the sequence has the following characteristics:

(1) The mercantile stage. The higher-income classes live in the city centre, in and around the seat of political power (as in London's twin city of Westminster). Usually they occupy one sector only, almost invariably that with the pleasantest physical environment.

(2) The onset of industrialization. As the city is transformed, with its workshops being replaced by factories and large numbers of immigrants being attracted, the congestion, pollution and perceived health hazards of the densely peopled inner areas increase considerably. The upper classes, who can afford the time and money, move to more spacious residences on the edge of the sector, distancing themselves from the negative externalities closer in. (They also remove themselves from the seats of power, but these are now firmly entrenched in economic hands and need not be guarded day and night.) Their former residences in the inner part of the sector are filtered down, or may be removed to allow the construction of non-residential buildings. Some of the inner-city residents may counter the negative externalities, however, and maintain their residences there, either by the activation of economic and political power (as, for example, in Mayfair) or because their possession of large tracts of land allows them to distance themselves from the negative externalities (as with royal palaces).

(3) Industrial expansion. As the city grows, so more industries take over land in the inner zone, forcing residents to seek homes elsewhere, along with the recent immigrants. The invasion and succession generated, both industrial and residential, poses threats to the higher-income groups. They retreat, producing migration patterns as inferred from the Burgess and Hoyt models. Meanwhile new, relatively cheap dwellings are being built for the lower-income workers in other sectors of the city, and suburbs are being developed for the burgeoning middle classes (the professional and managerial supporters of capitalism). This process may continue for some time, so that each sector has a zonal pattern in which the most desirable residential neighbourhoods are those on the city's edge; it can be observed in reconstructions of the growth of cities in many parts of the world.

(4) A halt by the rich. At some date the rich cease to move out further within their sector, in any large number, for a variety of reasons. Industrial expansion in the inner zones may have halted, for example, so that there is no further advance of their negative externalities. The rich may then decide to resist the invasions of the lower classes, enforcing this by a

combination of economic power (the pricing mechanism for homes) and political influence; the latter is variously expressed in restrictive covenants on land use and subdivision, in zoning schemes, and in the activities of estate agents. The preparedness of the rich to take this step may depend on their ties with the CBD (as industries decentralize – see next chapter – so the main employment opportunities of the city centre are in the offices inhabited by the white-collar workers), on the quality of the environment which they currently occupy and on the cultural connotations of certain areas. Thus in some cities the rich have been forced to move a long way from the city centre to avoid pollution; in others, they have stopped fairly close to the city centre, as in many of the post-industrial cities (of the American West, for example, and of Australia and New Zealand) which have little large-scale, noxious industry in and around the CBD. In several European countries, some of the rich have traditionally preferred inner-city homes, separated as much as possible from industries; many of these people also have second homes for week-end and vacation use.

(5) The suburban boom. In late industrial capitalist societies the relative affluence of many households, the relative cheapness of personal transport by car and the halting of the filtering process by the higher-income classes have combined to stimulate a suburban boom for the middle classes. (Despite the end of filtering and the outward migration of the highest-status neighbourhoods, the higher-income groups periodically improve their housing either by upgrading internally that in which they live – this option is common if the dwellings have architectural merit – or by replacing it *in situ*.) The availability of mortgages; the requirements of the construction industry; the building of large suburban public housing estates; and the spread of squatter slums of hope: all of these have generated suburbanization, often at considerable costs to the inhabitants.

The spatial pattern of suburbanization continues the sectoral form of earlier stages. The high-income classes, formerly living on the edge of their sector, are encapsulated by 'upper-middle-income' developments, studded perhaps by occasional higher-income nodes which may be focused on former exurban villages. Similar patterns occur in the middle-income sector. Beyond the low-income sector's edge, squatter settlements are likely in Third World cities, and council housing estates in Britain. Elsewhere new forms of suburbia are developing to house those who want their own homes but cannot afford a large deposit. In the United States trailer-home developments, comprising factory-built mobile homes, cater for this group, who come at the lowest-income end of the housing class comprising purchasers of homes on mortgages; in parts of Australia it is possible to buy half-homes, in the hope that, like the squatters, extra income later will make extensions possible.

This evolutionary sequence is only a very brief sketch intended to capture the general changes in patterns. It incorporates the situation in most contemporary cities, and suggests that the Burgess and Hoyt representations reflect just one stage in the process in one place. During the 1920s and 1930s in the United States only the most affluent could afford home ownership and the commuting times and costs involved with suburbanization. Since then a much wider spectrum of the higher social classes has been admitted to the ownership-based housing classes so that the outer zones of American cities are now the homes of the middle classes, who are segregated among the municipalities and neighbourhoods of suburbia by occupation, by income, by race and by age. The same is true of most other capitalist cities, in some of which the middle classes include squatter settlement residents, whereas in others they are the tenants of the newest public housing estates. Their relative positions in both society and space are similar, and they illustrate the obsolescence of the Chicago-based models.

Culture and urban residential patterns

These models of urban residential patterns were developed in a particular cultural context – the midwestern USA city in the 1920s and 1930s. Land there was widely considered as a commodity to be traded and used for maximum financial gain, and holdings were usually small. Thus there were few impediments to the outward extension of the urban area, and most landowners were prepared to sell their agricultural holdings to speculators and developers at prices well above the value of the land if it were to remain in agriculture. The United States was plentifully endowed with farming land, so a profit could be made and a new holding bought further out.

Some American critics of these models pointed out that the market forces were not the only determinants of urban land use. Sometimes, they argued, land is not sold for its open-market price, but is retained in a 'less efficient' use because this is considered socially desirable. In Boston, for example, the Common could have realized major profits if sold for commercial development, as could the sites of churches and other buildings: the cultural significance of these uses over-rode the market forces, however. Similarly, the residents of certain residential areas – such as the affluent on Boston's Beacon Hill – developed ties to their home environments which led them to prevent the incursion of other uses, whatever the price offered; the forces of invasion and succession were repelled.

Such cultural variables were accepted as possible causes of 'deviations' from the model patterns, but were largely ignored. In any case, they were relatively rare and the American free-market in land prevailed unencumbered by such cultural constraints. This was not the case everywhere, however. In Britain, for example, there are substantial estates owned by the aristocracy and their descendants and many of these people were not keen to release their land for urban development. As a consequence, large tracts of land close to the city centre were not released for residential and other urban uses, thereby constraining the evolving urban form: in Sheffield, for example, the south-southwestern sector was 'frozen' by its owner, the Duke of Norfolk, until the 1960s, when it was released for public housing; and in Nottingham the reluctance of landowners to release land all around the city contained urban development into a small area for several decades, leading to very high density, low quality slum housing for the rapidly-expanding working class. Thus factors other than market forces strongly influenced the social morphology of those cities. Further, when the estate-owners decided to release their lands they were able – through the leasehold tenure system – to influence the ways in which they were used: Sheffield again provides an example, in the controls on housing type and design imposed by the Earl Fitzwilliam on a large tract of land in the western sector.

It is for reasons such as these that the models of urban residential patterns receive relatively little attention in this book. Although the processes of distancing, congregation and segregation are general and identifiable in all capitalist countries, the actual pattern in any place reflects the local cultural context. As is argued throughout this book, the processes are general but their empirical outcomes are unique.

The townscape

All of this discussion has been about the social morphology of cities, suggesting that these are mosaics of separate districts characterized by the nature of their residents – their socio-economic class position, their housing tenure, and so on. This social morphology is both reflected in the built environment – the townscape – and is further subdivided by the varying appearance of the separate areas. Thus, for example, council housing areas in British cities have a number of identifying characteristics – including the scale of development and the uniformity of the housing within an area. (With sales of these properties to sitting tenants in recent years, some variety in the detail of the townscape has been introduced – in the colour schemes of doors and window-frames, for example – as the new owners express their individuality.) But within most cities many different types of council housing can be found – the terrace and semi-detached housing of the early decades, the high-rise blocks of the 1950s and 1960s, and the variety of low-rise developments in the 1970s.

The uniformity of council housing tracts in Britain is paralleled by the developments of state housing in East European cities. But it must not be assumed that housing provided by the state necessarily has these characteristics. In New Zealand, the state housing provided from 1935 on comprised a number of detached bungalow designs and estates were organized so that no adjacent dwellings were of the same appearance. Again, it is necessary to take account of the local context. The same is true of the privately-built housing for the working classes in the nineteenth century. English cities are characterized by the long terraces of two-storey by-law brick houses, for example (with inter-city variations reflecting the stringency of controls on back-to-back homes), but Scottish cities have tenement blocks and in Australia the small size of the average construction firm meant that there was considerable variety in housing types and styles along a single street. Within countries, too, there are particular local housing styles, such as the 'Tyneside flats' of the Newcastle area.

In general, the higher the status of an area – at the time of its initial development – the lower the density of housing, the larger the homes, and the greater the variety of styles. Even where 'estates' have been developed, they are usually small and uniformity of appearance has been avoided; in any case, since most have gardens, variety is rapidly introduced by the tastes of the owners. In general, the higher status areas also have less rigid street layouts. Working-class areas, and others developed as relatively large estates, tend to have either rigid rectangular street patterns or carefully-designed geometrical layouts; in some countries, the pre-urban cadaster creates discontinuities in these patterns, because the housing is fitted into the previous set of fields. In high status areas, on the other hand, a much less rigid arrangement of streets is common, in part reflecting the lower densities, the higher prices (which residents are prepared and able to pay: in lower status areas, the need for the developers to be price-conscious ensures that they produce 'efficient' street patterns which allow as much land for housing as possible), and often the piecemeal development process.

The townscape is a visible statement of the social geography of a city, therefore, reflecting the general processes of distancing and the particular local provision of housing and environment. Much of the townscape develops in a piecemeal fashion as landowners and developers react to the pressures of demand. This was common in pre-industrial cities, which were usually densely-settled and where processes of infilling on vacant land – much of it behind the original buildings on the main streets – were common. But in some cities, the details of the townscape have been the result of a conscious grand design: Haussmann's reorganization of central Paris for Louis-Napoleon from 1853 illustrates this, as does the 'city beautiful' movement in the United States, launched in Chicago in 1890. (Note, too, Engels' claim that the main thoroughfares of Manchester were designed to hide the slums from the rich commuters who lived on the edge of the built-up area, many of them in walled-in parks with gates at all entrances.)

The social geography of the city can be seen in its concrete manifestations, therefore, as well as in maps of who lives where derived from census and other data. Those concrete manifestations have in many cases been created both to aid the display of status, power and affluence and to maintain the social relations on which those characteristics are built.

Living in the city

The social geography of the city and the townscape not only represent the economic, social, ethnic, political and cultural relationships within society, but also form a major element of the environment within which those relationships are reproduced. Distancing allows people to ignore those they wish to avoid contact with, and to develop images of society that are inconsistent with the empirical reality: as Michael Harrington noted for the United States, spatial segregation ensures that many millions of suburban middle-class residents have no contact with the poverty of those penned in the inner cities, and makes it easy for them to assume that America is an affluent society.

People are socialized through their local environments. Thus segregation by class and race ensures that most individuals live their daily lives – especially in their childhoods, when most life-long attitudes are formed – among people similar to themselves. This can stimulate the development of local communities and class/racial cohesion and consciousness. But because such communities are localized there is no overall cohesion but rather the development of a mosaic of separate groupings. Politically, this suits the interests of the minority affluent groups, because it divides the working-class majority and makes concerted action against the status quo unlikely; groups fight local conflicts over their own bailiwicks rather than general conflicts against the capitalist system as a whole.

It is within these local environments that people live out their everyday lives, especially their everyday social lives. They view the city not as a series of planning zones, nor as quantitatively-defined social areas, nor as townscape components, but as environments for living in. They create images of those environments, mental maps within which they organize their activities; certain routes may be favoured, for example, because they are perceived as 'safe'. Such images are personal and group representations, a large series of social geographies in which the residents develop a sense of place and definition of themselves and of others.

Because people operate in these separate lifeworlds, their views of what lies beyond their defined boundaries are based on, at best, partial information. 'The world outside' is occupied by different people, with characteristics that make them distinct. Very often, those characteristics are expressed in negative terms, so that residents of other areas are perceived as threats. A 'them and us' situation is created, based on labelling and stereotyping rather than on contact. To counter these threats, community boundaries are defined; these are usually mental constructions only, but they are clear images and transgressions are seen as potential sources of danger – by children, by youth gangs, and by adults too. The everyday worlds may not have the legal characteristics of sovereign states, but to many people the reality of a road as a boundary is much clearer, and the fear of those on the other side greater, than is a distant international frontier. Urban life is spatially segmented, by those who experience and create it as well as by those who study it.

Conclusions

This chapter began with a discussion of the nature of urban social areas and the degree of class segregation between them. The focus throughout has been on places within the city and on differences between areas.

Address is important in capitalist cities (and in some non-capitalist cities too, at least for a portion of the population), as an indicator of social status, as a source of externalities and as an investment. But the co-ordinates of the map of social areas are social and economic, not latitudinal and longitudinal, which is why most of the discussion of the social geography of the city here has lacked references to relative locations in traditional mapped space.

The main theme of the exposition here has been a need to understand how capitalist society creates housing classes, and how the members of the various classes are allocated to different areas to fulfil the distancing requirements, to maintain the circulation of capital and to preserve inter-class economic differentials. The results of these allocation processes can be mapped as social areas and their pattern can be analysed as examples of certain spatial forms. But understanding the social morphology of the city requires an understanding of economic as well as of social processes, and without the former the latter can provide hypothesis-generating descriptions only. Such descriptions are very common in the urban geographical literature. The format of this chapter indicates that to understand that which they describe urban geographers must direct their attention to the underlying economic mechanisms of the societies that they study. The spatial morphology is not irrelevant, but it must be recognized (as Figure 7.5 makes clear) as the result of a complex of economic and social forces and not the reflection, as some geographers have argued, of the operation of an independent, spatial variable.

8. Non-Residential Land-Use Patterns

The previous two chapters have focused on the residential areas of towns and cities, with but a few passing references to all other land uses. This concentration can be justified because residential areas generally occupy the largest portion of an urban area and their homes are the centres of the residents' lives. Nevertheless, few towns are predominantly residential in character and it is the non-residential areas, especially those containing the industrial and commercial functions, which are both *raison d'être* and focus of urban life.

The focal role of non-residential functions refers not only to their provision of the urban economic base but also to their centrality within the town's life. The latter is reflected in movement patterns: a majority of journeys either begin or end at a non-residential land use. As such the location of those uses is crucial in the arrangement of the town. Indeed, in locational terms the residential areas may be considered a residual component, occupying that land not required for other uses. This comes about because in a capitalist city with a free market in property land is occupied by the highest bidder and in most cases non-residential land users can outbid residential users for a desired plot. Both a shopkeeper and a householder may favour a particular location, for example, the former because of the sales potential there and the latter because of its general location. In financial terms the tangible benefits to the shopkeeper are usually the greater, and so he or she can afford to offer more for the site. In addition, whereas householders have to meet housing costs out of their income, shopkeepers (like most non-residential land users) can translate their costs into prices, so passing on the expenditure on the site, which is usually a very small percentage of their total cost structure, to their customers.

The various types of land use within the city are interdependent, and create externalities for each other. In general accessibility is a positive externality, since it reduces transport costs: thus accessibility to place of work may be valuable to householders (as argued in Chapter 7) because it reduces their time and money spent on commuting. But proximity to workplace may contain a strong negative externality component, because of the noise, the traffic congestion and the air pollution associated with industrial land uses. Other non-residential land uses offer similar spatial patterns of benefits: accessibility to shops, schools, churches and parks may all be considered desirable by a household and worth paying extra for, but proximity (i.e., close accessibility) may be undesirable (because of the noise from schools and parks, for example).

Land users also provide positive externalities to others of the same type, as already illustrated for residential uses. Few factories are entirely independent of others; most get at least some of their input from other factories and dispatch at least some of their output to industrial users. Accessibility makes for the ease of transfer of goods and services between linked plants. Shops feed off each other in a similar way: a butcher, for example, may attract consumers from those visiting a grocer nearby, and vice versa. Others do not benefit from proximity, however; there is little to be gained from siting one park close to another, for example. As in the study of residential land-use patterns, the investigation of non-residential uses in the city must pay particular attention to the important role of externalities as an influence on location decisions.

Initial location patterns

There was relatively little spatial separation of the various urban functions in the towns of the mercantile era. Industries, shops, offices and residences coexisted in high-density built-up areas and two or more functions often shared the one building. (One of these functions was almost invariably residential, for most shops and workshops were operated from their owners' homes.) This absence of spatial separation reflected two features of mercantile society: the lack of mobility, which made journeys of even a few miles very time-consuming, and the small scale of almost all industrial and commercial operations.

The only exceptions to the above description concerned the functions involved with the exercise of both spiritual and temporal power within the mercantile settlements. Ecclesiastical land uses were generally separated from the commercial and industrial, as were the political and military, although in both cases the areas housing the élite classes also provided for several of the other land uses involved in servicing the powerful.

With urban growth and the development of industrial capitalism, the complex inter-mixing of land uses was slowly replaced by the spatial separation now common. This began with, for example, different types of workshop – each still with residential accommodation attached – occupying separate parts of the urban fabric. This establishment of separate quarters allowed the entrepreneurs to benefit from proximity to competitors and to linked establishments, and in many cases was a spatial statement of their oligopoly in their particular trade.

As cities grew, their centres became more congested and less attractive as residential areas: health hazards increased with the growing densities, for example. Thus the richer, capitalist classes sought to escape having to live among the noise and squalor, and divorced residence from workplace. The availability of horse-drawn transport allowed this separation and the initiation of suburbanization, and security permitted movement away from the source of their wealth. Their employees (other than domestic servants) no longer lived with them; they too had to commute (unless accommodation was available above their workplace), usually by foot over short distances from the rental housing now being provided. Thus a combination of positive externalities for industrial and other uses leading to agglomeration economies with negative externalities stimulating suburbanization for the rich initiated the evolution of a clear spatial arrangement of land uses.

This picture of towns and cities at the beginning of the industrial era suggests that, in the largest at least, there was a mappable pattern of land uses. London was an obvious example. Temporal power was concentrated at the western end of the built-up area, focused on the palaces at Whitehall and Westminster. Industrial and commercial land uses were further east, centred on the City. Within this latter area functions were spatially segregated, with many of them focused on a particular institution, such as the fish market at Billingsgate, the meat market at Smithfield and the fruit and vegetable market at Covent Garden. Some workers lived in the homes of their employers, but most of the latter had moved to the city's edge, mainly the hilly land to the north, whereas the former were confined to the rookeries and other low-quality residential areas of the East End. But cities of London's size were few at this stage of development and most lacked a detailed pattern of land-use separation. The urban growth of the industrial capitalist era was the main stimulus to spatial segregation.

Industries in the city

The development of the factory had a major impact on the spatial organization of urban land uses. The common image of factories is of large assembly lines employing several hundred people. Some of those established in the early industrial period, notably in textiles, had such characteristics, but most were much smaller, employing only a few workers. Dominance by large factories, characteristic of many modern cities, occurred only slowly.

From workshop to factory

The factories in the initial decades of industrialization were the descendants of the mercantile workshops and they retained many of the characteristics of the latter. Although each was self-contained, it was rarely involved in a full production process. Instead, it specialized in one part so that its output moved to another small factory for the next stage of the manufacturing process. Organization of the flow varied. In some cases the unfinished products were sold each time, each factory buying its requirements. In others the different factories acted as subcontractors to an entrepreneur who was moving the materials around and who was responsible for the sale of the finished good. In some, factories subcontracted for part of the work to be conducted by others (including, in some industries, those who worked at home).

Many industries were characterized by these organizational forms, which were reflected on the ground in clusters of linked small factories; this was typical of the manufacture of clothing, for example, and of furniture, jewellery, many small items of machinery and equipment (such as guns), and also of the printing trade. Many of the factories began in homes, or as annexes to dwellings, and their expansion eventually forced the residents to move elsewhere. Success in some cases led to the construction of a purpose-built factory, but most firms operated in premises which could easily be converted from one use to another. Thus the factory areas of inner cities were usually chaotic assemblages of small firms operating in tiny premises, often in poor physical condition and providing far from ideal working environments. They attracted other functions to the area, such as transport companies, warehouses and the manufacturers of related products, such as packaging materials.

As these small firms grew, so they expanded into other premises, either adjacent to those already occupied or in a separate part of the industrial area, where a larger site was available. Some of the firms stayed in the same family for several generations, maintaining traditional skills and continuing an independent line. A few succeeded such that they expanded and became public companies, increasing their volume of production manyfold and in some cases taking over linked functions via vertical integration. The expansion necessitated a move to new premises in other parts of the city for some; for others, growth was possible alongside their existing premises. The latter introduced a further element to the complex organization of the inner-city zone, introducing custom-built large factories to the congeries of small workshops and convertible factory buildings.

Many of the firms established in this inner area did not survive for long, and their premises were soon vacated, to be taken over by others, including a new generation of would-be industrialists. The flexibility of the small factory buildings meant that the inner area acted as a seed-bed for industrial entrepreneurship. Would-be factory owners could establish themselves in small premises with relatively little initial outlay and for low rents. If they succeeded they could either expand their premises or move to a better location; if they failed they would not be left with a large capital encumbrance. Thus the inner-city industrial area has been the growth point for many new enterprises (as well as the graveyard for many failed attempts); in Third World countries, the street-front plays a similar role for some industries still. With the evolution of late capitalism, and its dominance by large firms, this seed-bed function has declined somewhat in importance, but it still continues in those industries which incorporate small operations and encourage enterprise.

A commonly found land use within the inner-city industrial zone is the warehouse. Again, few of these have occupied custom-built premises in the past (the exceptions are mainly the largest firms); most occupy buildings that could easily be adapted for any warehouse, and probably a factory too. In some places the warehouses predated the factories in the area, and perhaps provided the catalyst for industrial development in the mercantile city. The preferred locations for many warehouses are close to transport terminals, notably ports and railways, which in some places produced segregation of the warehousing and industrial functions into separate parts of the inner area.

The picture painted in this section, therefore, is of a collar of workshops, warehouses and small factories surrounding the city centre proper (which is itself dominated by commercial land uses). Much of this collar may have been developed originally as a residential zone, with industrial and other functions being conducted in the dwellings. The expansion of non-residential uses created negative externalities which forced out some of the inhabitants – the rich, who could afford moves to the suburbs. Eventually, industrial and other expansion required the working-class residents to vacate their homes too, many of which were converted for industrial use. Thus the collar became dominantly non-residential, and residential developments were encouraged further from the city centre. As the industries and warehouses prospered and more were established, so the collar expanded, invading the adjacent residential areas and eventually ousting the population. But over time the small factory became outmoded, and it was the large establishments elsewhere in the city that came to dominate the spatial organization.

The assembly-line factory

Industrial expansion is accompanied by a concentration of ownership in its various sectors. The exploitation of economies of scale that this concentration involves means that small workshops are replaced by large assembly-line factories and semi-skilled, specialized machine operation replaces more extensive craftsmanship (based on a long period of apprenticeship) as the dominant function of the workforce. Factories are large because of the large volume of throughput necessary to their operations and also because the assembly-line procedure, by which parts are successively added to the frame as it passes along a conveyor, requires large areas of space. Associated with this, space is reserved outside the factory for expansion and for the storage of both inputs and outputs.

These large factories have much more voracious appetites for land than did the small plants of the inner-city industrial collar. Their need has been for open land in large tracts, undeveloped for urban uses, and relatively cheap. This was only obtainable on the urban fringe, so that factory industrialization on a large scale has been an important component in the spread of the built-up area. The distribution of factories has not been random around the urban edge, however. It has been clustered in certain sectors of the urban fringe, for a variety of reasons.

(1) Although large assembly-line factories are more independent than the inner-city small factories and workshops, they are involved nevertheless in a complex web of linkages involving products flowing from one to another. Clustering of such linked activities minimizes the time and cost involved in movement, reduces the volume of stocks each firm must hold (which ties up capital), and allows for external economies in the shared use of suppliers and service establishments.

(2) Almost all industries require an infrastructure, including basic utilities such as roads, water, power and the disposal of wastes. These are provided by external agents, either local governments and their subcontractors or the land developers, both of whom maximize their returns by concentrating their supply in certain areas.

(3) Industries are dependent on a transport infrastructure and prefer

sites where access to these is provided. Ports, rivers and canals are the initial foci in some cities; railways were important in the industrial development of many during the nineteenth and early twentieth centuries; today, road transport is predominant for the movement of most commodities in most countries and so access to the main fast roads is often vital to a firm. For a few, an international airport, such as Heathrow in London and O'Hare in Chicago, is a focus for their operations.

(4) Factories are widely regarded as sources of negative externalities for residential areas, such as noise, traffic congestion and air and water pollution. Thus town planning powers have been used to exclude them from most residential areas and confine them to a few segments of the outer zones of the city where their negative externalities are either minimized or so distributed that only a few people – the relatively poor and powerless – suffer impact (downwind of a pollution source, perhaps). A few areas, on the other hand, welcome industries to their neighbourhood, as potential contributors to the local tax base: this is the case with certain suburban municipalities in the United States (see p. 182).

(5) Although industry under late capitalism is dominated by a few large firms and giant factories, there are still many small establishments. Many of these depend on the larger units for their orders – they may specialize in subcontracting for small volumes of certain components, for example. They benefit from proximity to their customers, and in many cities are catered for by the provision of small premises on an industrial estate close to the big assembly-lines. Some of these small firms occupy flexible premises, which in part duplicate the seed-bed function of the inner industrial collar.

Industries outside the central city tend to be segregated on specially designed industrial estates and parks, therefore, with ample space and with access to the major transport modes. Development of such estates has been encouraged by planning procedures, and by the investment activities of governments attempting to attract industries to certain areas of a country.

The industrial pattern, the residential pattern, and commuting

The industrial land-use pattern of most towns and cities thus comprises two main elements: an inner collar dominated by small factories and warehouses, and a set of suburban clusters made up of large, usually purpose-built assembly-line factories and warehouses plus an aureole of dependent small premises. All of these clusters are not on the edge of the built-up area, of course, for later suburban residential and other developments enclose many. Most of the clusters are focused on a particular transport route producing, in some cases, a linear industrial belt.

The relative importance of these two components varies from city to city. Few have no industrial collar at all, but this may be very confined. Small central places usually have only a few inner-city industries, for example, reflecting the fact that their local markets are insufficient to provide many viable business opportunities. Among larger settlements, some mercantile cities in former colonies also lack an extensive inner collar because there was little industrial development in their initial growth and when industrialization was introduced it was based on imported assembly-line methods which required suburban concentration. In other colonial cities, especially those with dual economies (p. 131), inner collars of small workshops are dominant elements of the industrial pattern. Finally, in most state capitalist cities which have few small businesses operated by independent entrepreneurs, the workshop zone is not very extensive.

Some towns, such as the textile centres of Lancashire and New England and many mining settlements, were based on only a few industrial establishments. If no other industries developed, these places lack an inner collar of any size, and the single industrial focus may not be centrally located. But if the town grew, its economic base was expanded and multiplier processes were generated; small workshops were likely as responses to a multiplicity of local demands. Even so, major industrial centres vary considerably in the size of their workshop component. Places whose nineteenth-century industries were dominated by a relatively small number of large factories, such as Pittsburgh and Manchester, tended to have fewer inner-city workshops than others, such as Birmingham and Boston, where the small firm was a more potent force for industrial growth.

Industrial areas are the foci for much of the daily intra-urban flow of goods and people, with commuters dominating the latter element. As the inner collar developed and workplace and residence were separated, so two types of commuting evolved. The first involved the higher-status workers – owners, managers and professional employees – who escaped the negative externalities of the inner areas, preferring to travel relatively long distances from the more salubrious, lower-density suburbs. The other involved the employees, forced to move out of the inner collar by industrial expansion (including expansion of associated transport facilities: many lower-class residences were destroyed in British cities to make way for the railway termini in the nineteenth century). They moved into adjacent areas where housing was being built for them to rent and where the white-collar workers sought to distance themselves from their blue-collar counterparts, if necessary paying more, both for housing and in commuting costs, in order to achieve their social goals. Journeys to work from these areas were generally short and cheap; many were made on foot.

With the concomitant developments of relatively cheap, rapid public transport and suburban housing markets for the lower paid, so the average commuting journey of the inner-city industrial worker was extended. The city centre was the focus of the new public transport systems and drew workers in from all areas served by the tram and train lines. At the same time, the more affluent classes could afford to move even further out, perhaps to separate settlements beyond the city's edge, thereby keeping their distance from the approaching hordes.

Many of the new transport termini were located in the industrial collar rather than in the city centre proper, so that the industries there were provided with a large labour-pool on which to draw. The newer suburban clusters were not so advantageously located, however, and in the larger cities many of them were virtually inaccessible to major segments of the population, since to reach them would mean the residents of other sectors crossing the city centre, which may involve one or more change of vehicle. Thus the success of such industrial clusters in obtaining workers depended on the parallel growth of residential areas nearby, although this constraint was later removed somewhat with the more widespread availability of first inter-suburban buses and then personal vehicle ownership. And so again two types of commuting evolved. The higher-status workers distanced themselves from the new industrial clusters, preferring either the established, more desirable residential areas (and the consequent commuting) or new developments accessible to, but separated from, the workplaces. For the blue-collar and lower white-collar classes, on the other hand, either housing had to be found nearby in order that they might accept a job in one of the suburban clusters or choice of jobs was constrained by accessibility patterns from existing homes. This created many problems for the female members of suburban blue-collar households; many wives and daughters had to travel long distances to the city centre for secretarial and other jobs, whereas those prepared to work in local factories could often be exploited by manufacturers enjoying a virtual labour-pool monopoly.

There are several major components to the commuting pattern in a large industrial city, therefore. Factories in the industrial collar around the inner city draw workers from the entire urban area, along the public transport lines, with perhaps a sectoral component in the very largest reflecting the existence of several transport termini. The suburban clusters, on the other hand, draw mainly on the local population, thus creating semi-independent nodes within the general flow pattern. For both areas, the affluent workers tend to travel longer distances, from wherever the more desirable residential districts happen to be.

The accessibility constraints on commuting have been relaxed somewhat with the spread of car ownership, although there are still large segments of the population, distinguished by age and sex as well as class, among whom car ownership is relatively rare. But inter-suburban journeys are often difficult because they cut across the main suburb–city centre traffic flow, which itself is the cause of much congestion (as is discussed in the next chapter). Thus in some larger cities the location of industry is a major constraint on residential location decision-making. This results in a housing market segmented by location as well as by type, tenure and price. For the relatively affluent the constraints are least important, as flexibility in working hours plus dependence on private transport allow a choice of 'desirable' residential areas, of which there are usually several within reach of any workplace. But among the lower paid the restrictions on choice may be severe. For those in the mortgage-based housing class, where to live can be strongly influenced by accessibility to work requirements; in turn, once a choice has been made, changing job may be difficult without also moving home, which may not be enjoyed. For those in other housing classes and with less locational flexibility, where they work may be almost predetermined by where they can live, especially for those without their own means of transport. This may mean long, expensive journeys, or it can give some employers a monopolistic position in the employment market. And for a few unemployment may be the consequence of having to live in an area with no local jobs and with no public transport links to places with vacancies.

The office sector

The processes of industrialization and urbanization are inextricably linked, and the role of manufacturing in the city's economic base is frequently stressed. And yet in many towns and cities, especially the largest and the smallest, manufacturing industries have never dominated in the provision of jobs. And today it is the tertiary and quaternary employment sectors, the shops and the offices, which are the major employers in a wide range of urban settlements.

Offices developed in the same way as industries in mercantile cities, as offshoots from city-centre commercial establishments. Some provided services for the general population; some were more specific in their markets, catering for certain other types of business establishment; others developed as the white-collar divisions of manufacturing firms. Initially all were small, but with the normal capitalist dynamic some expanded, thus increasing both their employment and their space demands. The range of office functions expanded too, as more varied services were required to maintain and manipulate the capitalist industrial system. The growth in office employment so created contained three main types of worker: decision-making and managerial; routine operations; and servicing (mainly secretarial).

One major growth pole within the office sector has been connected with the operations of the state. The role of the state in the initial development of industrial capitalism was confined to the provision of protective functions only, such as law, order and defence. These made little demands on the decision-makers (the politicians) and only small support staffs were needed. But as the state's role expanded, incorporating both provision of a bulwark for capitalists and a guardian of the working class's living standards, so the size of its workforce has grown, requiring employees of all three types. And in state capitalist societies, the excision of the private office sector has produced a monolithic state bureaucratic system.

Offices, like factories, vary in their dependence on others. The headquarters office of a large industrial firm, for example, may have sufficient business to justify employing its own legal staff; even so, it will be reliant on other firms for certain specialized legal services. Most offices depend on a large number of others for a wide range of services that they cannot provide for themselves. As a result, offices tend to cluster together. Some of this clustering reflects the links between different office types – there are bank branches throughout office areas of most cities, for example. Others reflect the links between offices of the same type, as with the many lawyers, most of whose business is conducted with other lawyers. And yet others reflect the links between certain offices and a particular institution, such as between barristers and the law courts, stockbrokers and the stock exchange, trading companies and Lloyd's of London, and so on. Thus there is not only a separate office land-use area within the city but also clear functional segregation within this area.

The office area of most cities is centrally located, adjacent to the retail sector and within the industrial collar. In many this reflects the inertia of initial locations, plus a range of more recent factors. Some offices need to be accessible to the general population, and so must be accessible to the major transport foci. Most must be accessible to their employees, the majority of whom are white-collar workers and live in the higher-status suburbs. Many of the routine and service workers in offices are not highly paid, however (this is especially the case with women workers), and travel to work by public transport; this constrains many offices, especially the larger ones, to a central location. Thirdly, the links between many offices and particular institutions means that the location of the latter dictates the siting of the former: the popularity of city-centre locations for seats of government and other major attractants maintains the demand for other types of office space nearby. Finally, some offices occupy prestige buildings and locations which, unlike the small workshops of the industrial collar, cannot be abandoned without major capital loss. The firms concerned are only a minority, but many of the other smaller firms are so closely linked with those occupying the prestige sites that they cannot afford to move away from the centre.

The prestige factor has been accentuated in the last few decades in many cities by a boom in office development. Most firms rent their office space, and the growth of that sector at a time of relative stagnation in the industrial sector in many countries made investment in office buildings very attractive for speculators and developers. There was a resultant boom in both the provision of offices and their rents, with many prestigious developments. (These were encouraged by local governments for the potential tax revenue that they offered.) Firms are to a considerable extent 'locked in' to accepting what is provided, so that the joint forces of developers and local planners in the promotion of central-city projects committed firms to locations there.

There has been much less decentralization of offices within urban areas than there has of factories, in part because concentration and centralization have not been as extensive in the office sector and in part because whereas modern factories need extensive areas of land, the space requirements of offices are readily met in prestigious skyscrapers. Large firms and government departments can operate their routine components away from the expensive premises of the city centre, however, with telecommunications allowing contact with headquarters. Thus there has been some establishment of office developments at major suburban nodes in the largest cities, providing that access for the workforce is satisfactory. Other firms have moved away from the city centre to occupy converted older large homes in desirable suburbs. But not only does the bulk of the office sector still reside in the city centre, it now dominates the provision of employment in the urban heart of the world's major cities.

Commuting patterns associated with offices are similar to those for factories: those in the city centre tend to draw on the whole urban area whereas those in suburban locations are more reliant on adjacent districts for a supply of workers, especially in the routine and service types. Compared to factories, the inner-city offices draw their employees from further afield, because of their greater demand for white-collar workers. The link between central city and outer suburbs that this involves has stimulated the provision of high-speed, relatively long-distance commuting modes, such as motorways and fast train services. These have encouraged further suburban expansion, either sprawl (as in the United States) or the 'colonization' of independent settlements beyond the built-up area (as in Britain) and have made some of the areas so connected closer to the city centre, in terms of travel time, than areas physically closer to the urban core. The construction of motorways into the city centre has also disrupted inner areas, both lower-status residential neighbourhoods and some industrial districts.

Although central-city offices obtain many of their workers from the suburbs, substantial numbers of their employees live close to the city centre, in the flat and apartment areas provided for the young and the childless. Until recently most of these were in the routine and service categories, except in the largest cities, but the recent gentrification trends (p. 207) are resulting in a greater concentration of workers of all types relatively close to the office concentrations. In absolute terms, however, such developments have had only a small impact on the volume of commuting from the outer suburbs with its immense pressures on limited transport systems for a few hours each day.

Retail land uses

Retailing is both a major employer in urban areas and a significant focus to the city's spatial organization. The majority of households send at least one member on a shopping trip at least weekly; most generate several such trips each week, and some generate more than one every day. Since most shopping involves the consumer travelling to the vendor, the location of shops is an important influence on the choice of which to patronize. The consequences of these choices are represented in the urban spatial pattern.

Central place hierarchies

Geographers have conducted many investigations of intra-urban retailing patterns and the related shopping trips. Initially, their work was structured within the context of central place theory (see Chapter 3), which was developed to account for the distribution of retail and related functions in an urban system but which offered profitable insights for the study of intra-urban shopping centres. Thus a hierarchy of centres within the city was postulated, focused on the central business district (CBD). This meant that in a town with a three-level hierarchy, households were assumed to obtain their daily needs from the nearest small (neighbourhood) centre and their weekly requirements from the nearest medium-sized, or district, centre; there were fewer of the latter than of the former. The less frequently purchased commodities (often termed comparison goods, to indicate the mode of shopping which involved comparing the products of several shops before deciding on the purchase; daily and weekly shopping was for convenience goods) were bought in the CBD, the single centre offering the desired commodities. Centres at each level in the hierarchy would contain all of the establishment types found in the smaller centres.

This notion of an intra-urban central place hierarchy is intuitively reasonable, and is in general format confirmed by casual observation. The CBD is almost invariably the largest shopping centre, containing types of establishment not present elsewhere within the urban area, and there is usually a plethora of small shopping centres providing consumer goods for their surrounding populations. Questioning shoppers suggests that many use the nearest outlet, except when they want to buy something available only at a distant centre at the same time as they are purchasing a widely sold convenience good; in such circumstances they may make a single, multiple-purpose trip to the distant centre rather than buy one commodity there and the other at their local neighbourhood centre. Thus in general there is both a hierarchical pattern of centres and a hierarchical pattern of shopping trips (or at least there has been until relatively recently).

Casual observation and simple questioning, which produce supportive evidence for the central place hierarchy notion, are contradicted somewhat by closer investigation. Many people do not shop at the nearest available centre selling the required goods, for example, and the pattern of centres is usually by no means as regular as the theory would suggest. (Figure 3.1 can be translated to the intra-urban situation to suggest what should be observed.) There are many reasons for these apparent deviations. One concerns urban densities. These vary considerably with distance from the city centre, as outlined in the previous chapter; higher-density areas can support more closely packed arrangements of shopping centres. Everywhere within the built-up area population densities are very much greater than in the rural areas for which central place theory was derived, however, and because of this urban dwellers have many more separate shopping opportunities available to them within a given distance of their homes than is true for rural residents. Many shoppers are not so sensitive to the time and cost of travelling to a centre other than the nearest that they are constrained to using the closest alternative, as is the case for those living in the areas for which the theory was devised. Moreover, whereas the area occupied by a central place system as perceived by the theory is fixed so that the only change involves the relative growth and decline of its settlements, the area occupied by a growing town is expanding and new shopping centres are being created within a matrix of existing nodes. The theory cannot easily cope with this. And finally, the theory makes unreal assumptions about the nature of retail businesses and about shoppers. For these reasons, the central place analogy has many drawbacks, and a developmental approach to understanding retail land-use patterns is preferred.

The development of shopping patterns

The current CBD of most towns and cities was both the original focus and the location of the urban economic base. Initially many of the urban functions operated from temporary premises (such as market stalls) on certain days of the week only, to be succeeded later by the establishment of permanent shops, facing onto or close to the market focus. Thus the main retail centre began as the focus of urban commercial life and the node towards which transport routes were oriented.

As towns grew, their populations attracted the establishment of more shops, of two types. The first were those which duplicated the existing establishments in terms of commodities sold, and which competed with the established firms for a portion of the expanding market. The second were those which initiated a widening of the range of commodities traded. These replaced the former practice of having either to 'import' the goods that they sold from other settlements or of doing without – perhaps making the required commodities in the home. Some of them represented a process of specialization whereby an establishment was set up to perform a function formerly available as part of the range offered by an existing store. The advantages of such specialization included wider choice, better service and particular knowledge, as well in many cases as lower prices.

In locational terms the new shopkeepers could operate one of two strategies. The first involves the search for a near-monopoly over a portion of the market, and was most apt for those setting up as competitors to established stores. The shopkeeper defined a target level of turnover, identified the necessary population that would produce this, and then sought a location which would be closer to that number of consumers than was any other store. Assuming that people bought from the nearest available outlet, he would then gain a monopoly over the population of his hinterland. The other strategy involves seeking as large a proportion of the total market as possible, which involves competing with like establishments. Analyses suggest that the best place from which to compete with a shop is close to it, so that potential customers can compare the two offerings. This is a strategy of locational interdependence, therefore, which involves the competitors being near-neighbours in the same centre. It can be combined with the first strategy if the aim is to compete not for the entire urban market, which presumably requires a location in the CBD, but only for a portion of it, which involves establishing stores in the district centres of the three-level hierarchy discussed above.

The first of these strategies produces a dispersed pattern of shops, with each at the centre of its exclusive market area. Logically, each would be at the centre of an area containing exactly the number of consumers necessary for a viable business, and only if two different shops (selling different goods) required exactly the same supporting population would shopping centres emerge. But the calculations cannot be that precise, and they are based on many uncertainties, such as the degree to which the local population will patronize their nearest store. Furthermore, some shops of different types benefit from proximity to each other, such as grocers and butchers, as well as from the reduced travel costs of visiting two nearby. Thus most of the dispersed shops, located according to the first strategy, are in small shopping centres where a few different types of establishment serve the local market. (Isolated shops are only common in very densely peopled areas, such as the working-class districts of nineteenth-century British cities where the front rooms of houses were converted into small establishments.)

The search for a spatial monopoly involves an element of risk, because it means forecasting an unpredictable future. An entrepreneur may establish a shop in a new suburban area, for example, where he has no local competitors. Further expansion of the built-up area beyond that which he serves may attract another shop, however, which may be closer to some of the original entrepreneur's customers than he now is. If some of these people transfer their custom to the new shop, the original one may lose so much turnover that it goes out of business. Development of a relatively large shopping centre at the second location could virtually ensure the failure. The likelihood of this happening is not always clear, and many entrepreneurs have taken the gamble in the past, and their businesses have survived for a reasonable number of years. But conservative instincts might suggest to the individual setting up on his own that the second strategy is the safest, that a shop in an established centre, to which people are accustomed to travel for a variety of goods, is a securer investment than a gamble on an unproved location. The result of this would be that many suburban areas containing sufficient purchasing power to support certain types of retail establishment in fact lack those shops, because no entrepreneur is prepared to take the gamble and invest his capital in opening a store there. Thus the pattern of shopping centres and of shops is more centralized than would be the case if central place theory applied.

Whereas the first strategy encourages a dispersal of shops through the urban area, the second leads to concentration in the CBD and the major suburban centres. (The exception would be the type of shop which served only one segment of the urban population – perhaps a certain class or age group – which is spatially segregated. Such shops, including the retailers of certain types of clothing, are best located close to their particular market.) It is the balance between the two strategies which accounts for the distribution of shops at any one time: the greater the concentration on the first strategy, the greater the dispersal of shops and the more important the role of the small shopping centre as a focus of urban life.

As the processes of duplicating and import-replacement proceed, in parallel with both urban growth and changes in consumer tastes and spending patterns, so the relative importance of the two strategies may change. Thus, for example, the first few shops in a town exclusively concerned with television rental probably clustered in the CBD, but as the demand for such stores increased so some may have been established in major suburban centres, seeking to serve only a portion of the market. In general, as duplication proceeds so the shopkeepers entering a particular sector are more likely to seek a location in a relatively small centre (relative, that is, to the size of centre in which such shops are currently found), so that the CBD remains very much the location for the most specialized and infrequently demanded stores only. The process may operate in reverse, of course, as the number of establishments in a certain functional category declines (as with men's hairdressers in recent years).

Recent changes in the retail pattern

The discussion so far, as is common with work dominated by the approach of central place theory, has assumed that all shops are operated as independent businesses, that each shop is the exact replica of all others selling the same merchandise (in terms of size, etc.) and that the aims of all shopkeepers are the same. This is far from the case. Some shopkeepers are the equivalent of peasants and are satisfied with a certain standard of living, but many are run as capitalist enterprises with the same drives as comparable manufacturing firms. The major aim of a capitalist shopkeeper is to increase profits, which can be achieved by either or both of increasing the turnover and profitability of his present shop and establishing a chain of shops.

Attempts to make the operation of shops more efficient have produced major changes in the nature of retailing in recent years, particularly in the retailing of foodstuffs and other convenience goods. A major alteration has been the replacement of labour costs by land costs, achieved by substituting self-service for counter service. This requires space, and to afford the land prices shopkeepers must either ensure high turnover or locate where land costs are low. For the latter alternative, the new types of stores have been located on the edges of built-up areas (or, in some cases, on derelict land such as unused industrial sites). For both, the need for large trade volumes has involved attracting shoppers from long distances, many of whom travel by car and buy in large quantities (perhaps only once every two or three weeks). Parking facilities are one of the attractions. Another is relatively low prices. These can be achieved not only through the economies from the change in selling style but also from the bulk-buying necessary for large sales.

The development of the supermarket for food retailing had a model in the earlier growth of department stores. These developed at the time when most shoppers were dependent on public transport and so did much of their comparison shopping either in the CBD or, in the biggest cities, in the main suburban centres. To succeed, these stores had to win custom from the specialist shops, which they attempted through a variety of strategies, some of which were based on price. (Again, bulk-buying for large-volume turnover was crucial: the largest were able either to establish their own manufacturing operations or to contract with manufacturers for particular types of merchandise carrying the retailer's own brand.) The growth dynamic involved the department stores expanding into new lines of merchandise, a strategy now adopted by the supermarket chains which have established their own brands not only in foodstuffs but also in a wide range of other goods, both convenience and, traditionally at least, comparison. Some have established superstores in suburban locations which trade in a great range of merchandise and which are serious competitors with established shopping centres. To compete with them, specialized retailers have established similar types of store, such as the large cash-and-carry furniture warehouses found not only in suburban locations but also on industrial estates.

In retailing as in manufacturing, therefore, the dynamic processes of capitalism have involved concentration and the growing hegemony of a small number of large firms. Some retailing sectors have been relatively untouched by this trend: the sale of jewellery has remained with small shops, for example, whereas that of many types of clothing has not. This has important locational consequences. The large stores are major magnets for customers; the smaller ones usually are not, but are dependent for much of their trade on the joint magnetism of their neighbours. The latter are what might be termed 'location decision-making sheep', following the lead of the larger firms on which they are partly parasitic. Thus the department stores and the supermarket chains are not only the most likely to pioneer new locations, they are also sought after by the developers of planned shopping centres. The restriction of commercial development to certain zoned areas, the large infrastructure of parking and other facilities necessary for a major suburban centre and the cost of land have combined to make development an attractive investment for building companies. To ensure their venture's success, the developers will seek to lease at least one store to a major outlet (either a department store chain or a supermarket chain). Development may not proceed beyond the initial design stage if the location cannot be sold to such a magnet. Once it has, other small firms may be attracted to take out leases in the belief that the major tenants will draw sufficient custom to the new development. The large firms, with chains of shops, can afford a reasonable gamble on a new site, financing it out of the profits earned elsewhere; the smaller firms, many of them with one or two shops only, cannot take such risks for they lack the cushions of capital and profits elsewhere.

There is a symbiosis between the large and small firms in retailing just as there is in manufacturing, because certain commodities and services generate insufficient turnover for the large firms to bother with them. Department stores and supermarkets concentrate on mass consumption goods with large and rapid turnover, therefore, whereas smaller shops can survive by concentrating on specialist items such as *haute couture* and certain types of bread. Other small shops survive because of the service they offer, which may be one or more of personal attention, access, and convenience of opening hours. Thus whereas the concentration and centralization trends in retailing are aimed, as in other sectors of the economy, at oligopoly (if not monopoly) they do not result in the total demise of the small store in a capitalist system. Indeed some of the specialist services of such stores increase in popularity as affluence increases, and it is mainly the shops that try to compete in the sales of mass consumption goods that suffer from competition with the large companies.

Prices and planning

Competition within each sector of the retailing industry is severe, most especially in the sale of foodstuffs in which price is the major weapon. The large supermarket chains can sell cheaply because of their economies of scale in buying and selling, and they offer a wide range of goods, often in a comfortable, weatherproof environment which means that shoppers drive into a covered car park and can obtain all they want under the one roof. The small shop can offer accessibility, in time and space (although many supermarkets open until late in the evening in some countries), but its goods are usually relatively expensive because of its small turnover and inefficient labour-intensive methods of selling. To meet the competition of their larger rivals some small firms have formed buying cooperatives which allow them the benefits of large-scale purchases but leave them with their independence as vendors (some advertise jointly, so that at least a proportion of their prices are common). Thus food retailing is characterized by three main types of shop. The supermarket chains are usually the cheapest for a standard basket of groceries and the independent one-shop firms are the most expensive. There are variations within each category, of course: among the stores operated by the big firms it is the largest which are often the cheapest of all because of their major economies of scale obtained both from very large turnover and the combination of retailing and wholesaling functions in the one place – the hypermarket – which cuts out the middleman.

Similar patterns of price variation are now available in many merchandising lines such as furniture and electrical goods: the large firms balance their mark-up (the difference between buying and selling prices) against their turnover, and achieve their desired profit levels by making a little on a lot. The small shops survive by making a lot on a little, which they attempt to supplement with the associated services that they provide. This price variation offers the shoppers a choice – either the cheap supermarket a car journey away or the expensive, local shop which will deliver, take orders and perhaps allow credit too. But for some consumers there is no choice. Many supermarkets are accessible only by car, except for those who live nearby, and they are located in the suburbs away from the inner-city homes of the relatively poor and not linked to those neighbourhoods by public transport. (Public transport is not very advantageous for shopping at supermarkets, since to benefit from the journey cheap goods must be bought in sufficient quantity to cover the travelling costs. Without a capacious car boot in which to carry those goods, the journey by bus is usually financially not worthwhile.) As car ownership spreads and the retailing industry becomes ever more oriented to the mobile shoppers, so the burdens of the immobile increase. Their small local shops become relatively more expensive, but even so they may generate insufficient trade to remain financially solvent. Closure of convenience goods stores is widespread now in both suburb and inner city, to the detriment of the poor, the old and the immobile who cannot travel to the distant, if cheap, supermarkets and planned shopping centres. And the CBD, which might still be accessible to them, is probably in decline too, as the big firms close their shops there and move to the more attractive outer suburban developments.

The location of new retail developments is now covered by town planning legislation in most capitalist cities. Within existing built-up areas, zoning for shops has largely involved recognizing the existing nucleations, with perhaps some allowance for expansion at a few and a realization that many of the smaller centres are in decline and should be abandoned. In newer areas, however, planning procedures could lead in creating the new retailing pattern by determining where shops would be located in the general public interest. But zoning can only indicate where retailing will be allowed; it cannot insist that it occur in certain places nor can it prescribe the types of shop that will be opened in its retail zones. The general result of several decades of planning has probably been a greater concentration of small shops in neighbourhood centres than would have been the case with complete freedom of choice for the entrepreneurs. Whether this is good or not is difficult to evaluate.

With regard to the major new developments, the planning process has been much more of a passive component of the development sequence. The large retailing and development firms involved in the design and finance of big suburban centres have considerable economic and political power, although some of their schemes have been successfully resisted by local residents' groups. Frequently they have been able to get plans changed in order to build on land that they own, or have taken an option on a site at a relatively low price reflecting its current zoning for, perhaps, residential usage. The developers rather than the planners are rewriting the retail pattern of the city in some capitalist countries, such as the United States and France. They have been less successful in others. In Britain, for example, the classes in political power in the inner cities have resisted suburban developments in order to protect the CBD (and they have in general not been competing with suburban municipalities with separate zoning powers, as in the USA). Thus most British cities have undertaken ambitious city-centre redevelopment schemes in recent decades, many of which involve office as well as retail premises and have been conducted by private development firms whose large profits have been guaranteed by the local government sponsors.

Within the planned suburban centres and in the city-centre redevelopments, the firms involved in their detailed planning have incorporated one feature of the earlier unplanned centres. Most large shopping centres, and especially the CBD, display considerable spatial regularity in the location of the various types of shop. Because it is the large department stores which attract most shoppers to the centres, these are usually at their foci. Around them are clustered the more specialist comparison goods stores which hope to attract passing customers, in part by their window displays. Beyond are the more specialized stores, to which most shoppers make specific trips and few are attracted by a passing whim. (Furniture stores are typical of this group. They are usually in clusters in case customers wish to compare offerings before making a decision on what may be an expensive purchase.) Other specialist stores, many catering for a relatively small market only, occupy an outer zone and the main thoroughfares into the centre. Within this segregated pattern are distributed convenience goods stores which are located according to the spatial monopoly strategy; examples are newsagents, chemists, tobacconists and food stores catering mainly to workers in the centre and its environs.

Much research has been done on the patterning of stores within centres (and also of departments within large stores) to identify the best spatial organization in terms of turnover and profit maximization. This is then put into operation in new developments, as part of the fight between suburban centre and CBD on the one hand and between city-centre redevelopment and the traditional shopping centre there on the other. The competition is not only for the urban area shoppers but also for the visitors from the hinterland, many of whom now find the suburban centres, with their convenient parking, more attractive than the CBD.

Indeed, where it has not been protected by the planners, the latter centre has rapidly declined in importance and status in some cities, increasingly operating only as a convenience goods centre for lunch-hour shoppers from the nearby offices.

Summary

The developmental sequence for urban retail patterns outlined here begins with periodic markets, usually only one per town, passes through a central-place-like system comprising a hierarchy of shopping centres, and ends with the emergence of a system of large planned shopping centres competing with, and in some cases defeating, the established centres of the CBD and the inner suburbs. There are vestiges of the earlier stages in most cities. Many British towns still have periodic markets, for example, many of them operating weekly on open sites and specializing in relatively cheap goods; others have permanent market halls housing similar establishments, and there are a few specialized street markets. And the supermarkets of the latest stage in the sequence have not stimulated the total demise of the small shop and the neighbourhood centre, although both of these aspects of the central place hierarchy era are in considerable decline as a consequence of the combined forces of concentration in the retailing industry and increased mobility for customers.

Cities have a dual, if not a triple, system of retail provision, therefore, with to a large extent each system being oriented towards a certain clientele. Dual systems are particularly apparent in the cities of the Third World, where they reflect the existence of dual economies (p. 131). In the large, ex-colonial cities of Asia and South America, for example, there is usually both a 'conventional' CBD serving the affluent élite and a bazaar or informal sector of street traders, many of whom operate without either licence or even temporary premises. And certain cities within plural societies have separate, often independent, shopping systems for each cultural group, exemplified by the adjacent white and Indian CBDs in Durban, and by the identifiably different facilities for blacks, for Jews and for various South European groups in American cities. Like other aspects of the urban pattern, these are changing as capitalist development proceeds, with the rate of change varying from district to district within the city.

Other non-residential uses

Housing, manufacturing industry, and shops and offices are the major land uses within towns and cities, and only two other important uses – parks and streets – have so far not been discussed. Other use categories can be identified. Most occupy only small amounts of urban land but their significance in daily life may be much greater. They are considered here under two headings – private and public service provision – to illustrate the basic factors underlying the relevant locational decisions.

Private service provision: Health care

Outside the state capitalist countries, the provision of several elements of health care is undertaken by private individuals, usually operating within constraints imposed by the public sector. Most important of these services is general practitioner medical care; dental services are also provided privately, but hospital facilities, for much of the population, are usually in the public sector.

The location of general practitioner services can affect the quality of health care in two ways, irrespective of any possible variations in charges made. First, the more accessible a doctor's surgery is to a patient's home the easier it is either for the patient to visit the doctor or for the latter to make a home visit. Second, the greater the density of doctors serving an area (or the ratio of doctors to population) the easier it should be for patients to obtain treatment. Ideally, therefore, doctors' surgeries should be evenly distributed through a city's residential areas, with perhaps some bias towards the neighbourhoods of greatest proven need for their services (which are probably the neighbourhoods with the greatest proportions of either young children or older people). But this rarely occurs, and most cities have a very uneven distribution of surgeries. In part this reflects an under-supply of doctors, because of careful control of their numbers by the medical profession (to ensure a sellers' market for their services) and because general practice has a low status within the profession and most graduates prefer hospital practice. In addition, doctors can choose where they practise. In countries where they charge the patients for consultations, as in the United States, only the altruistic choose to establish surgeries in other than relatively affluent neighbourhoods where the patients are adequately insured and meet their bills. In those where the doctors are paid by the state, they apparently prefer to live and practise among their peers, so that even without a financial incentive (British doctors are paid according to the number of patients registered with them) the higher-status residential areas generally have much better doctor/patient ratios than do the lower-class districts. Thus in Britain the greatest density of doctors' surgeries is in the older high-status neighbourhoods, relatively close to the city centre, where doctors operate surgeries from their homes. Suburban areas, and especially the large public housing estates, are very poorly served, with consequential inter-class differences in health care.

There is a recent trend in many countries for the concentration of general practices into fewer, larger units. Until this began, most practices involved between one and three doctors operating from perhaps two locations. The newer group practices have been introduced to produce economies of scale in the provision of para-medical services, with perhaps as many as ten doctors operating from the single surgery which contains a range of allied functions for diagnosis and treatment. These health centres serve large populations, and their introduction reduces the general level of patient accessibility to doctors, since the average person now lives further from a surgery. Since research has shown a definite correlation between the use of medical services and the distance travelled, this can result in some lowering of general health standards (assuming that the frequency of consultation is a reasonable surrogate for the need for medical services). Where the health centres are located can influence which groups suffer most from this aspect of concentration and centralization, therefore; as with the changing distributions of jobs and of shopping opportunities, it may well be the poor and the immobile who are most disadvantaged.

Public service provision

Local governments are responsible for providing a wide range of services for their territories. Some of these are locationally specific, such as parks and recreation centres, hospitals, libraries and schools, police, fire and ambulance stations; others are general, such as streets, or involve taking the service to the homes, as with sewerage removal and garbage collection. In an ideal situation all homes should be provided with the latter, whereas for the former variations in accessibility to services should be kept to a minimum. Such ideals are rarely achieved.

Even within the constraints imposed by the number of locations that can be served, the provision of public services is rarely equitable. Such services are usually termed impure public goods, therefore, to indicate that they are not equally provided for all. With some services, variations between neighbourhoods in both quality and quantity are not easily accounted for: that some districts have better-maintained streets than others, more parks or better libraries, appears to result from nothing more than an apparent random allocation of public funds. Some distributions can be related to the age of neighbourhoods: parks are much more common in new suburbs than in the older, high-density parts of cities, many of which were built over before legislation was passed giving local governments the power to impose statutory minimal areas of recreation space per district.

Variations in the provision of some services can be accounted for by investigation of the relative power of different neighbourhood groups within the civic system. Cases have been taken through the American courts, for example, alleging that certain local governments have discriminated against black residential areas in the provision of services such as garbage collection and street maintenance. Indeed, charges of political favouritism are common in most towns and cities from time to time, as powerful groups in control of the local government machine favour their own residential neighbourhoods in the allocation of desired services or distribute public spending as political largesse to win votes in important constituencies. The opposite pattern is frequently charged too, of politically powerful groups distributing negative externalities so as to avoid their own neighbourhoods. These externalities include major developments, such as intra-urban motorways, and localized facilities, such as community centres for released mental patients and prisoners, which local residents do not want in their streets. Frequently the local government is challenged by residents' groups on such schemes, but not always with much success.

Charges of neighbourhood favouritism and of locational decisions taken to favour certain groups or to court electoral support are difficult to prove. Almost invariably, a public authority has insufficient resources to meet all the demands made upon it and so must decide between what are all seen as perfectly justifiable claims. Rationales can be produced to support all locational allocations, and arguments advanced that in the long run distributional equity will be ensured. And in some countries the balkanization of local government allows some groups to escape from contributing to the demands for public services made by others less fortunate than themselves, even to the extent that they freeride on the provision made by and for the latter.

Conclusions

Urban areas are organized around their non-residential land uses, for these provide employment and the places where the inhabitants spend their money. In order that all urban residents might be treated equally, these land uses should be distributed through the urban fabric in an efficient pattern. But the definition of efficiency as it applies to the consumer (which is usually that the desired land use should be easily accessible) is not necessarily that of the producer. For most non-residential land uses the producer is a capitalist operating a business and seeking profits. His location decisions, and the combination of all those taken by others like him, may well produce a distribution of, for example, shops or factories which in terms of access (and perhaps other criteria too, such as negative externalities in neighbourhoods) is much more beneficial to some classes in society than to others. The involvement of the state in creating these distributions does little to alter the situation, for the planning process is closely allied to capitalist desires and in many cases has promoted the sort of spatial distribution which the firms want. Thus, as the major land uses discussed in this chapter are the focal points of much of urban life, their spatial distributions, and therefore the spatial organization of urban life, ensure that the capitalist ethic is strongly etched upon the changing urban fabric.

9. Rearranging the City

The modern city is the focus of much analysis by commentators of various types who point out the range and intensity of social and economic problems concentrated there. To a considerable extent such concentration is not surprising. The larger part of the population of most countries lives in a few large cities, and it follows that social and economic problems are likely to be concentrated there too. Thus many problems identified in cities cannot properly be termed problems *of* the city. They are in effect problems *in* the city, reflecting the concentration of population that results from the social and economic processes that also generate the problems. Other problems are more precisely *of* the city, in that they are consequences of population concentration rather than symptoms of it.

The analytical separation of 'problems in the city' from 'problems of the city' has not been recognized by many commentators, who assume that all problems are of the latter type and so can be solved by some kind of urban planning. Like the problems of urban systems identified in Chapter 5, many of the proposed and enacted planning solutions are either ameliorative problem-solving or allocative trend-modifying (p. 125). Discussion of such planning involves an analysis of its probable success in removing the identified problems, therefore, many of which are located in the spatial organization of the city.

City problems

A wide range of urban problems has been identified and solutions attempted. For ease of presentation and discussion here, they are categorized into five groups.

(1) The problems of the urban environment, of which the main examples are air and water pollution. To some extent these are problems in the city since, for example, certain activities will produce the same volume of pollutants wherever they are located. But they are also problems of the city because agglomeration of the major sources of pollu-

tants into certain places leads to concentrations of their outputs whose joint effect may be greater than the sum of their separate impacts.

Air pollution is the product of three main generators: industrial plants; domestic fires; and the internal combustion engine. Their relative importance varies from time to time and place to place, depending on the types of activities in the place, various aspects of its climate (cold weather means more fuel being burned, for example), microclimatic factors and the density of traffic. (Combination of high traffic densities with temperature inversions may cause a very nauseous pollution problem for example, as in Lima.) Water pollution also emanates from both industrial and domestic sources and their emission of waste products. The disposal of solid wastes is a further major environmental problem.

Urban areas are significant modifiers of local climates, through the heat they generate and their alteration of the land surface and its cover. Most places generate a heat island effect under certain climatic conditions. This is most pronounced over the densely peopled and the industrial districts, with the consequence that city centres are sometimes several degrees warmer than are the lower-density suburbs, whereas the latter are warmer than adjacent rural areas. In itself this is rarely a problem, and neither is the general tendency for cities to be somewhat cloudier and less windy than their rural environs. At a local level, however, the climatic differences may be magnified by the orientation of buildings, which can have a considerable effect on wind speed and turbulence.

Some of the climatic effects of urbanization result from the replacement of a natural vegetation cover by buildings and paved areas. These also influence the hydrological cycle. Instead of rainwater percolating through the soil into streams, as is general in 'naturally vegetated' areas, most of it falls on impervious surfaces and runs straight into the drains and natural water channels. This rapidity of run-off means that urban drains, streams and rivers have to cope with much greater peak flows during periods of heavy rain than is the case in rural areas where, apart from exceptional circumstances, rainwater only slowly gravitates towards the streams. To cater for this difference, the greater the proportion of the land surface that is built over, the larger must the drains be to cope with the run-off during outbreaks of sustained heavy rainfall.

(2) The problems of urban spatial organization. Cities reflect the complex division of labour which means that individual firms and households can only be maintained by a great volume of interaction. The movement of messages can be organized along cables and through the air, but most of the interaction involves the movement of goods and

people. This requires both a substantial communications network and a transport system that can cope with very heavy demands.

The volume of interaction has increased more rapidly than has the urban population in most cities, representing the increasing complexity of urban economic and social life. Furthermore, it has become subdivided into a relatively large number of separate journeys as a consequence of increased private vehicle ownership. Few parts of the intra-urban road network were created to carry the present traffic volumes, particularly in and around the city centre during the two peak travelling periods each day. Despite the resulting congestion and frustration, however, alternative transport systems have not proved generally successful. Public transport loses passengers to the private car. It puts up its fares in consequence, to cover its overheads, and reduces its services; as a result it loses even more custom. In some of the biggest cities the road system is completely unable to cope with the traffic and so public transport services are maintained through subsidies.

As the organization of cities becomes more closely attuned to the needs of private car owners, so problems arise in the provision of a transport service for those lacking access to a vehicle. Many of the latter cannot afford their own vehicles; others are unable to operate them. For them travel is difficult and expensive, and many destinations cannot be reached. With the spatial concentration of many of the facilities which such people may want to use (as described in the last chapter) their quality of life may be seriously impaired.

(3) The problems of the social environment, a group which encompasses a wide range of issues. Many of them are associated with what commentators see as a decline of community within urban society, a consequent rise of individualism and a general decline in the standards and quality of social life.

The model for much of this concern is the perceived community-focused social and economic life of rural and small-town settlements. Such a focus is believed to have characterized urban areas in the past, with each neighbourhood having a high degree of community identification and containing a great volume of social interaction. But the city now lacks such localized communities and most of its interaction comprises impersonal economic contacts between individuals. The expansion of TV ownership in recent decades has accentuated this perceived collapse of urban social organization.

The lack of community in urban neighbourhoods is seen as the cause of a number of other problems, including those associated with interpersonal conflict. Cities tend to have high crime rates, against both people and property. The distribution of such crimes in part reflects the distribution of opportunities (shopbreaking is not surprisingly concentrated in certain areas). But there is a clear spatial pattern to both the occurrence of many types of crime and the residences of convicted criminals: the worst areas, it is argued, are those where community ties are absent and social cohesion is at its weakest.

(4) The problems of deprivation in the inner city. Virtually every capitalist city, and probably most non-capitalist cities too, contains chronically unemployed, even unemployable, people, desperately poor families and severe problems of personal circumstances. In most places these are concentrated in particular areas, especially those in the inner zones where housing is old and of poor quality. In recent years the existence of these pockets of economic and social stress has been widely recognized, in part because of a general increase in the number of poor and unemployed reflecting the current problems of capitalism. Thus academics, governments and social commentators have identified what they term the inner-city problem, which reflects the concentration there of the relatively deprived – those lacking jobs, satisfactory incomes, substantial housing, adequate educational facilities, and so on.

(5) The problems of urban government. The increased role of the state in the development of late capitalism means that governments, especially local governments, are deeply involved in the amelioration of urban problems and in the search for lasting solutions to them. Government efforts to revitalize the city by themselves involve problems, however. Many are unable to meet the full gamut of demands made on them because the expense would not be acceptable to those who meet the cost – the local taxpayers. Indeed some cities, such as Cleveland and New York in the USA, were virtually bankrupt by the late 1970s. And the general problems of finance are in many places exacerbated by the balkanization of some urban areas into a near-anarchic chaos of overlapping separate authorities.

These five types of problem are not independent, and treating them as such in part conceals both the complexity and the enormity of the current difficulties in urban areas. But presentation requires that they be discussed separately, a course of action supported by the fact that their treatment in many places indeed assumes that they are independent and that separate solutions are feasible.

Environmental problems

Although they are far from completely solved, of the five categories of problems just identified it is in tackling the environmental problems that progress has been greatest. This is probably because both the problems themselves and the solutions developed have no major social implications; they are engineering problems requiring engineering solutions. There are many economic implications, of course, which affect all people, although rarely equally.

Pollution has been tackled at its sources. Air pollution from domestic and industrial premises results from the emission of the waste gases from burning certain fuels. The worst source of such pollution is coal, but its use has declined substantially in recent decades with the availability and cheapness of cleaner fuels, including 'refined' coals. Probably the major change has been brought about by the banning of open fires in homes and of untreated emissions from factories. The British Clean Air Act has resulted in manyfold improvements of the clarity and odour of the air in cities. Smogs are now extremely rare and diseases related to noxious air are much less common than they were two or three decades ago. (The extent of the 'clean-up' is very apparent to those flying over South African cities. There coal and wood are still burned in the black townships, producing a pall of smoke which contrasts markedly with the clearer air over the low-density white suburbs where other types of fuel are used, especially for cooking.) Some of the replacement fuels produce their own problems, such as the removal of nuclear wastes, but because production of the energy usually takes place outside the main cities, this is not perceived as a specifically urban problem.

The problems of water pollution have been tackled by insisting on the treatment of wastes prior to their discharge into natural waters. Industries must purify the waters they emit and must not put toxic materials into streams, or even into catchments; domestic wastes are collected and treated prior to their entry into natural waters. These policies replace the often cavalier dumping of untreated wastes into rivers, lakes and seas, and reduce the threat to life in those waters. The effects are widely visible: many rivers in large cities are much cleaner than they were only a few years ago, and support fish again (as does the Thames at London).

Treatment of wastes rather than their emission into water and air has had the effect of increasing the volume of solid wastes which must be disposed of in urban areas. This contributes to an already increasing volume resulting from the rapid obsolescence of many consumer goods ('throw-away' capitalism) plus the volume of packaging materials in which such goods are sold. The amount of waste material to be removed in a large city is enormous, as is clear when those who remove it go on strike. Some of it is toxic, of course, whereas the deterioration – even just the existence – of other types can generate health risks. Domestic waste alone in the United States includes more than 30 million tons of paper, 100 million tyres, 30 billion bottles and 60 billion cans annually, plus several million cars.

The commonest treatment of solid wastes has been to dump them at a site where the negative externalities to neighbouring uses are low. (Such externalities include smell, the attraction of birds, mammals and insects, and the possible toxification, even eutrophication, of waters.) In the past most towns and cities have had convenient disused quarries or other sites where such wastes could be dumped but the increasing volume being produced has meant that many large settlements have found the provision of dumps very difficult. Alternative treatments have been sought, therefore, including compression of the wastes to a small fraction of their former volume, destruction through incineration, and recycling. All involve expense, which may in some cases be recoverable, and highlight the problems of servicing an advanced capitalist economy.

Finally there is the pollution created by internal combustion engines. The exhaust fumes are not only nauseous but also potentially toxic, with the level of toxicity reflecting the use of certain additives, such as lead. Other than either reduction of traffic volumes or their re-routing away from residential areas whose inhabitants may suffer the toxic effects, the main solutions involve improvements to the engines (reducing the volume of exhausts) and changing the constituents of the fuel; both are expensive. Alterations to engines are also necessary to minimize another form of pollution from traffic, noise, although where this is intense, as around airports, buildings require soundproofing to provide adequate protection.

Solutions to most of these problems are available, but many of them are costly. For a number, the initial costs of meeting anti-pollution standards are met by the producers, but unless they are either prepared to forego profits or can cover the costs through improved productivity they will pass these costs on in prices. Thus the consumers pay, as they do if some of the costs of pollution control are met out of government grants. Clearly the argument is that the reduction and eventual removal of pollution is for the good of society as a whole and that an affluent society should be able to afford such measures; others claim that survival of life on earth depends on undertaking such measures. But there are problems in allocating the costs. If producers pass them on in price rises, then it is the relatively poor who carry the largest burden, especially if the costs impinge on food and other basic commodities, which they invariably do. Because of the operation of the housing market with respect to negative externalities, the poor usually benefit least from anti-pollution measures, at least in the short term. If governments meet the bills out of a progressive tax system, then the payment is more equitable. Without governments, certainly, individual firms are unlikely to take many pollution-reducing initiatives on their own, as applications of the prisoner's dilemma game (p. 136) suggest.

Problems of spatial organization

The spatial reorganization of cities, involving the allocation of different land uses to separate areas, providing the communications network and coordinating the flows of vehicles and people, is generally associated with the practice of town planning. Such planning has a long history, for many pre-industrial towns were designed and redesigned *in toto* by their rulers and some industrial settlements result from the decisions of a few developers rather than a large number of uncoordinated individual locational decisions. But overall control over the expansion and internal organization of large capitalist cities was weak until the last few decades; the main constraint on what was placed where was price, allied with the desires of landowners, with perhaps some control imposed by the providers of certain utilities, notably piped water.

Town planning as a widely practised profession, and as a requirement of local governments in many countries, has precedents, therefore, but its main tool, land-use zoning, has been generally used for only a few decades. The stimulus for its widespread adoption was the increasing problem of negative externalities and the wish of those likely to be affected by them to protect both their quality of life and their property values. Zoning proceeds in most places through the definition of categories of land use and the preparation of maps which show where each can be located. It is permissive in procedure in that each zone on a map indicates what uses will be allowed there, but the planners have no power to ensure that the land is used for those purposes. Usually there is a scale of values, which may run: (1) industry, (2) warehousing, (3) retail, (4) residential, and any zone in a certain category is also allowed to include uses in a higher (larger-number) category. Thus an area zoned (3) is intended for retail uses; industry and warehousing are prohibited (except that such uses already present may be allowed to remain though may not be redeveloped), but residential uses may be developed.

Zoning indicates what the planners see as the desirable spatial organization of the city, therefore, but it is not certain that this goal will be achieved. The main problem is not that uses excluded from certain areas are located there, for this usually involves a full public inquiry at which involved individuals may present their case before a re-zoning is allowed. Rather it is because the zoning is indicative rather than directive and the desired uses may never be located in certain areas. The direction implied in a zoning plan is most likely to succeed if few alternative sites are available for the relevant land uses. Relative to projected demand, the supply of land has to be equivalent to needs. But this creates other problems in capitalist cities, for such a restriction of supply relative to demand fuels speculation in land and inflation of its price, with consequent costs for the potential users. The normative element of planning by zoning faces problems where the use of land is determined by the private market rather than by the planners.

A major development in the role of planners since the Second World War has concerned the construction of road networks that can cope with the rapid increases in vehicular traffic. For this task a methodology was developed for predicting likely future traffic flows based on observations of the traffic-generation of current land uses and of trip origin and destination links. Assumptions about the future growth of traffic allow them to predict where the new demand will be greatest, and assumptions about changes in the land-use arrangement allow a determination of where this increased traffic will be moving between. Such predictions allow the preparation of plans incorporating new road systems.

The technology of this planning methodology has become considerably sophisticated in recent years, providing the ability to produce more detailed forecasts of traffic volumes whose validity is entirely dependent on the assumptions regarding future traffic generation by various land uses and future vehicle use. From these forecasting devices models have been developed which produce the most 'efficient' land-use pattern/transport system, with efficiency being defined in terms of accessibility and total transport costs. Plans for new towns and for major urban expansions have been based on such models.

These plans and the assumptions on which they are based have been the focus of many attacks in recent years, especially by the residents of neighbourhoods through which motorways and other major road developments have been planned. Local protest groups have in some places been buttressed by the expertise of national organizations formed to resist the extension of motorway networks and their negative impact on residential communities. In some cases these groups have succeeded in stopping a development: several North American cities have unfinished motorway networks with, as in Toronto's Spadina Expressway, incomplete links testifying to the political successes of protest movements. But many of these roads have been constructed. They are important for the movement of white-collar employees from the outer suburbs to their offices in the CBD, replacing the railways and having the same effect as those earlier routeways on the inner-city residential areas of the lower-income groups. To cater for this traffic, in particular with respect to parking, requires comprehensive redevelopment of the city centre. In countries whose cities have a substantial history, this needed redevelopment threatens the architectural heritage and provides a major source of dispute between those wishing to protect the environment and those wanting a road system to accommodate the increased volume of traffic.

Public transport is still widely used in the largest cities by commuters, but in many cases the transport system is run at a loss. This is because the peaking of movement in the two rush-hours requires a large rolling stock which must lie idle for much of the day, thereby creating fixed costs which are not easily passed on in high fares. Introduction of fare increases usually results in a decline in patronage, which must be countered by either further fare increases or a reduction in service if the transport system is not to be run at a loss. Compared to road users, rail transport systems often operate at a disadvantage. The rail system must not run at a loss even if it does not make a profit, so that its costs must be met out of fares. But few road users pay tolls. Instead, roads are paid for out of the general exchequer, although some sources of revenue such as petrol and vehicle taxes may be available for spending on roads only. To some extent both the private motorist and the bus companies can externalize their costs, therefore, which is of particular benefit to freeriders who live outside the territory of the administrative unit that maintains the roads they use.

The problems of public transport are not confined to fares: the changing spatial organization of the city also makes for great difficulties. Efficient public transport services, which get their passengers to their destinations rapidly, are dependent on a high density of traffic generation, both at the origin (the suburbs in the case of commuters) and at the destination (the workplaces). If such densities do not exist, then the potential traffic for any one route does not justify its operation; this is particularly the case with railways, of course. The increasing decentralization of employment described in the previous chapter plus the development of large tracts of low-density suburbia means that there is insufficient demand to justify links between many workplaces and the range of suburbs on which they draw for employees. This forces commuters onto the roads in their own vehicles and adds to the pressures for road improvements which are being resisted by the protest lobbies.

Although the planning of land-use patterns by zoning and the provision of transport routes to allow commuting and other movements is part of the same general task, linking the two is often not easy. This is in part because of the 'indicative only' nature of planning already discussed, and in part because the planners provide only the routes, in the case of roads, and not the services that operate on them as well. Further, in many cities the administrative division of the built-up area into several independent planning authorities makes for difficulties in co-ordination. Perhaps the biggest problem, however, is that the utopian conceptions of normative planners who produce blueprints for the city of the future are resisted by the reality of having to fit their plans into an existing urban fabric; they are not given *carte blanche* but are tightly constrained by the existing situation. Further, their grand designs involve massive construction programmes which few authorities are able to afford except over a very long period. The staging of the implementation often reflects ameliorative problem-solving, therefore, with immediate traffic problems being tackled rather than the grand design being pursued. This often has the effect of merely shifting the problem (down the road, perhaps) and deflecting interest from the overall design.

The ethos of urban planning in capitalist cities is based on normative concepts of spatial efficiency, therefore, but the aims of spatial reorganization and redesign are often frustrated by the problems of working in a constraining environment. The grand design often has to make way for piecemeal attempts to cope with immediate problems and is harassed by financial and administrative problems, plus the vested interests involved in backing and opposing the proposed changes. Planning proceeds slowly, therefore, and the aims of one plan are often made obsolete by developments during its long period of implementation. The relative increase in the price of petrol is one such potential frustrating force for the planners in the near future, which may cause a major reconsideration of the desirable form of the city of the future. Meanwhile, many developments may proceed which conflict with the likely content of the next plan and whose implementation could seriously disadvantage those affected by them (such as the residents of new outer suburbs who become marooned because of the high costs of travelling to the city centre).

These problems of urban planning in capitalist societies should not be encountered in state capitalist and socialist cities which lack the private property market and which grant complete control over the land decision-making processes to the state bureaucracy. Thus ideal city arrangements based on notions of spatial efficiency should be feasible, as in Russia where there are planning norms indicating the maximum journey-to-work times which should exist in cities of different sizes. But, as pointed out in Chapter 7, the division of the state capitalist system among the bureaucracy there has separated industrial location decision-making from decisions about housing provision by the local soviet. The result is frequently a mismatch between the two and consequential long journeys to work for at least some workers. Without a monolithic single planning agency responsible for all aspects of land-use decision-making, planning seems fated to be at best only partially successful and the spatial organization of the city far from efficient.

Communities and neighbourhoods

Urban life is characterized by conflict. Some of this takes place in the political arenas, where individuals and groups fight the proposals of planners, of other bureaucrats and of individual firms, proposals which are perceived as potential sources of negative externalities and environmental deterioration. Other conflict is less institutionalized. It also involves groups within society and frequently has spatial implications. In parts of cities, for example, local street gangs define their 'turfs', perhaps marking the boundaries with graffiti; incursions by members of other groups are resisted, sometimes physically, so that inter-group conflict comprises a series of raids and pseudo-raids across the territorial boundaries, and success for one group might allow an expansion of its sovereignty. In other parts of the city groups of very different types may engage in the same basic activity, defining their territories and resisting incursions with a variety of procedures including, in some cases, recourse to the courts (see Chapter 5). Often the groups are latent, coming into formal existence only when a threat is perceived.

The city is a mosaic of territories, therefore, many of which are dominated by the members of a particular class or group and which have known, if not identified, boundaries. Some of the territories are exclusive and, as shown in Chapters 6 and 7, their definition reflects the social and economic processes which are typical of capitalist society. For some, the territories exist to protect class differentiation; for others, like the street gangs whose turfs overlap the territories of other groups living in the same areas, they are marks of status and social solidarity.

Some commentators believe that the existence of this territorial mosaic accentuates class conflict within urban society. It is a common human trait, they argue, to shun the unknown and avoid social contact when one's reception is not certain. Thus groups and classes distance themselves from each other out of fear based largely on ignorance. Distancing reduces contact and the ignorance which it fosters leads to the development of stereotypes which are usually gross caricatures (perhaps even total misrepresentations) of the characteristics of the groups with which there is no contact. These stereotypes accentuate the desire to avoid contact, since they suggest that major problems and negative externalities will result from such intercourse. Thus the territorial boundaries are stoutly defended. This set of processes may create particular problems for those confined to certain territories only, such as the racial ghettos of many cities. The stress may generate frustration (perhaps focused on perceived exploiters from outside the ghetto, such as shopkeepers) which occasionally erupts into violence and conflict across the boundaries.

How might this conflict be reduced, and if possible removed? To some, the answer lies in reduction of the levels of inter-group ignorance. The grossness of the stereotypes must be replaced by information based on close contact between the conflicting groups and classes. Contact will lead to a realization that the differences are not as gross as the stereotypes suggest, and this in its turn will promote inter-group harmony. Social contact requires residential mixing, and so to remove the conflict spatial segregation must be ended; the consequence will be inter-group tolerance.

The neighbourhood effect

The argument described in the previous paragraph is typical of a more general case made by social scientists and others regarding the role of social environment as an influence on individual attitudes and behaviour. This role is commonly termed the neighbourhood effect.

Attitudes and behaviour patterns are learned, based on observation and attempted emulation of models; this process is termed socialization. As argued in Chapter 6, the key period for socialization is childhood, and a major element in the process is the school. Most children attend a school near to their home, in many cases selected by parents because its models are acceptable and the attitudes, values and behaviour of most of its children and their parents are consonant with their own.

What happens when a school contains more than one model in substantial numbers? Research suggests that in a school with a mixture of, for example, children from middle-class and working-class homes, then the children in the minority class tend to adopt the norms of the majority. These two classes have very different attitudes towards education in some urban societies. The middle class are very much achievement-oriented and stress the value of educational success in the search for career and adult security. The working class are more apathetic, however, preferring that their children leave school and enter the workforce as early as possible. (That the working class, or some groups within it, do not operate the 'educational success brings income' syndrome is in the interest of the middle class since it reduces the competition for the latter's jobs and maintains a poorly trained under-class for the more menial occupations.) And so in a dominantly middle-class school the ethos is pro-education and working-class children in a minority in such schools tend to adopt the general attitude. Furthermore, their educational performance usually benefits too, for such minority working-class pupils in middle-class schools tend to achieve better examination results than do their contemporaries in dominantly working-class schools. (The reverse occurs as well; middle-class children in a minority in working-class schools aspire to less and achieve less than do middle-class children in middle class schools.) Most schools have some mixture of pupils in terms of class origins, of course, because few catchment areas are entirely homogeneous and the processes of residential segregation rarely produce exclusive suburbs. Relatively few have a sizeable minority from one class, however, unless the catchment area is deliberately designed to produce such a feature.

The validity of the research findings regarding the effects of school composition on pupil performance is the focus of much social-scientific debate. Other variables than class mixture may be the key determinants, such as the nature of the educational resource, as indexed, for example, by teacher quality and expenditure on textbooks. Nevertheless strong arguments have been made that a mixture of social classes in schools will be to the benefit of society as a whole, since it is more likely to realize the latent abilities of all children than is a class-segregated pattern. Such a case is one of the foundations of the British comprehensive school policy. But the problems of operating such a policy are many. For it to work, the children from working-class backgrounds must be in a clear minority in each school, which is a statistical impossibility in most cities; and the middle class of course have no desire for their children to be in the minority in a working-class school. And how might such integration be achieved, given the existence of residential separation of the classes? Busing is possible, and is employed for the mixture of racial groups, but with this the mixing is in the school only and the children return to their separate ghettos each night.

Education is not the only field in which the neighbourhood effect is believed to operate. Many attitudes of individuals and of groups, concerned with a wide variety of topics such as race and religion, political parties and particular elections, are influenced by their contacts with the views of others. As a result, for example, some social scientists and politicians argue that good race relationships will be fostered and the course of assimilation advanced if immigrants are residentially mixed with their host society rather than separated out into ghettos; the policy so proposed is sometimes known as 'pepperpotting'. The same course has been argued with regard to religious groups, as in Northern Ireland: experiments there have suggested that inter-group contact does reduce the animosity based on fear and ignorance but it is doubtful whether such small-scale experiments could be repeated in a volume sufficient to defeat the overwhelming prejudice.

Criminal activity is frequently related to the social environment also. Many parents claim that delinquent behaviour by their children reflects their 'falling into bad company' in the neighbourhood, and adults too can be led astray by persuasive models. Group criminality, such as football crowd hooliganism, is an example of this effect, of individuals behaving in company as they would not when alone.

Given the apparent pervasiveness of neighbourhood effects, therefore, it is not surprising that their operation should be encouraged in ways which are perceived by those making the case to be beneficial to society as a whole. Thus, for example, the crime rate might be reduced by mixing potential bad elements with a majority of good; race relations can be improved by residential mixing and a reduction in inter-group ignorance; educational achievement can be improved; and so on. But the case is in almost every situation based on circumstantial evidence only and the processes underlying the neighbourhood effect in its many guises are far from understood. What is the threshold for the effect? If a 10 per cent criminal element is introduced to a law-abiding neighbourhood, will the law-breakers renounce their former ways, whereas if 11 per cent is introduced is it the law-abiding element who are converted? Without answers any social planning (spatial engineering, as it is cften known) based on the effect will be more a statement of faith than the implementation of a proven process. And of course it assumes that the forces of evil already exist; in the above example on crime, what produces the 10 or 11 per cent criminal element, and would it not be better to try and remove criminality at its source rather than cure it after it has broken out?

Balanced neighbourhoods

The arguments regarding the neighbourhood effect and its potential inclusion in social planning are based on social-scientific work. A more general argument, along the same lines, is advocacy of the balanced neighbourhood, a concept which is deeply ingrained in certain planning ideologies; as an operational idea it is at least a century old, having been advanced by the Cadbury family in its plans for a residential development at Bournville. The basis for the concept is that there should be a mixture in each neighbourhood of all relevant groups within the urban society – classes, ages, races, religions, etc. – roughly in proportion to the composition of the society as a whole.

There is an extensive literature on the potential benefits for balanced neighbourhoods. According to Wendy Sarkissian, nine major arguments have been advanced.

(1) The mix will raise the standards of the 'lower classes' by encouraging them to emulate the attitudes and behaviour of their 'social superiors'.
(2) Aesthetic diversity will be encouraged and standards in general will be raised.
(3) Cultural cross-fertilization will be advanced.
(4) Equality of opportunity will be promoted.
(5) Social tensions will be reduced and societal harmony increased.
(6) Competition, even conflict, between individuals will be promoted, thereby fostering individual and social maturity.
(7) Operation of the city will be advanced by providing the lower-class population with middle-class local leaders.
(8) The stability of residential areas will be promoted, since undesirable 'invasions' will not occur.
(9) The diversity of the modern world will be introduced to every small community and neighbourhood.

Many of these arguments parallel those based on the neighbourhood effect.

There have been some attempts at putting the balanced neighbourhood concept into operation, particularly with regard to the housing of immigrant groups. A genuine balance has rarely been aimed for, however, and even less frequently achieved. Some private residential developments have attempted to introduce a range of family types, but there have been difficulties in selling or renting such homes. In British planning the notion of the neighbourhood, derived from earlier American practice, was the basis for the development of large suburban estates and the new towns built after the Second World War, but the neighbourhood was defined in terms of population numbers only, and although some efforts were made to instil community consciousness, little in the way of social mix was attempted. Where it was, in some of the new towns, it met with resistance from those members of the middle class in particular who preferred separation; in at least one new town – Corby – whereas the working class employed at the large steel works were housed in the town, the middle class were provided homes in nearby villages.

It is extremely difficult to introduce balanced neighbourhoods in capitalist cities, where all of the economic and social forces operate against them. The members of the owner-occupier housing classes, in particular, almost invariably indicate their belief in the neighbourhood effect model and seek to distance themselves from other classes, using the price mechanism. But in any case the concept of balance and its beneficial effects on society is a hypothesis only, and one which lacks empirical support. The fundamental assumption is that proximity generates social contact, but experience in a state capitalist society such as Russia, which has class differentials, clearly illustrates the fallacy of this assumption. The mixture of professionals and members of the working class in apartment blocks produced by the housing allocation procedures has not fostered social contact between the two: the strata create their own separate communities. Groups can be mixed physically, but not socially, and with freedom of choice in social contact class differentials (and age, race, etc.) are likely to prevail.

There are other problems with regard to implementation of the balanced neighbourhood concept. Will the working class agree to accept middle-class local leadership, or will politicization of the former put the two groups in conflict? Studies in villages 'invaded' by middle-class commuters suggest the latter. And will the relatively poor enjoy trying to emulate their more affluent neighbours? They could be frustrated by their inability to do so, on financial grounds, with their frustration fuelled by the demands of their children. The balanced neighbourhood concept has within it strong assumptions regarding a unitary, classless society lacking cultural differences, but such a society cannot be achieved by spatial engineering alone. While social and economic divisions remain, and people wish to retain them, the balanced neighbourhood will be resisted; if imposed, the members of the various groups will ignore it as far as possible. Conflict across territorial boundaries may be removed

– if migration to create new exclusive communities is forbidden – but it will reappear in another guise. As with all planning based on ameliorative problem-solving approaches, problems and conflicts which are inherent in a particular form of social and economic organization will not disappear with spatial reorganization; they merely reappear somehow and somewhere else.

The cycle of poverty and the cycle of affluence

Many of urban society's major economic and social problems are concentrated in the inner-city areas, as indicated earlier in this chapter. Realization of this fact during the last decade has been accompanied by much research into the problems of identifying the extent of deprivation and of providing remedies. Most of the policies proposed aim not at the individual sufferers of deprivation but at their environments; they are part of the general belief that social planning can be accomplished through spatial planning.

The cycle of poverty

A number of ways by which the emergence of inner-city problems might be understood have been proposed. That favoured here is the cycle of poverty (Figure 9.1) which locates the problems in society as a whole but indicates why they are spatially identifiable.

As its name implies, the cycle of poverty involves a continuous process which transmits relative poverty from one generation to another and makes escape from deprivation very difficult. In describing its operation, the most convenient point at which to start is the poverty itself. This, in the vast majority of cases, results not from any action of the individuals concerned themselves but rather from the operation of reward-allocation procedures in a capitalist society. Certain occupational groups are accorded only low incomes, which make for low standards of living; many members of these occupation groups are more likely to suffer from unemployment than are their contemporaries in other groups.

Low incomes generally mean that households cannot afford good-quality housing; that which can be afforded usually has to be lived in at high densities, with much sharing of rooms, and yet relative to incomes the rents may be high. The environment may be an unhealthy one, and the stresses and strains produced by poor health may accentuate those induced by having to live in a state of relative poverty. Poor physical

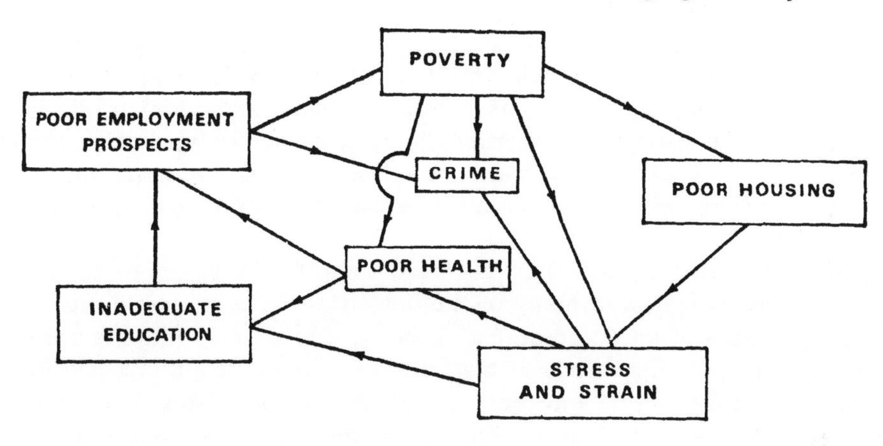

Figure 9.1 The cycle of poverty.

health may result from the combination of one or both of a bad diet, a consequence of poverty, and insanitary housing; poor mental health may reflect the strains of coping with such conditions. Together these make the pursuit and retention of employment difficult, accentuating the problems of poverty.

The stresses of poor housing and poor health will have an effect on the educational prospects of children born in such conditions. The home background will encourage early entry into employment, because it brings money to the household, even though this puts short-term needs before longer-term potentials; the physical environment of the home will not encourage personal study; the school may lack the learning resources to provide a good educational environment; and the models in the local environment do not encourage any alternative perspective. The end-result is inadequate preparation for employment in a technologically advanced society, poor job prospects and low incomes. Job prospects are not currently being advanced by the general crisis of capitalist production and the movement of firms seeking unskilled and semi-skilled workers away from the city centres and into suburbia and small towns.

Several of the features of the cycle of poverty identified in Figure 9.1 combine to encourage criminal activity. Crimes against property may reflect the problems of poverty itself plus the idleness resulting from unemployment. Crimes against the person, on the other hand, may result from the stresses of living the cycle of poverty, which may be accentuated by housing conditions, especially those relating to crowding.

A large number of studies have been undertaken which cull data from a variety of sources to illustrate how the cycle of poverty is concentrated

spatially, especially in the inner-city areas. Such concentration results from the operations of housing markets which were discussed earlier, plus the changing distribution of economic activity. Those in the cycle of poverty are constrained to certain housing classes only, so that the mechanisms of spatial segregation confine them to particular areas, which is where the indices of multiple social and economic deprivation locate them.

The problems of the inner city have come to the forefront of the public consciousness in many of the industrialized countries in recent years, because of the accentuation of the cycle of poverty with the deep economic crisis. In periods of high unemployment, people trapped in that cycle suffer most, and their spatial segregation means that those sufferers are physically concentrated. Furthermore, it is in the inner city areas that many of the largest factory closures have taken place, as companies restructuring their operations in the face of intense competition for contracting markets seek more docile, less unionized, and cheaper labour forces in other places. Many inner-city areas in Britain in the early 1980s had unemployment rates of 30 per cent or more: the young were especially hard hit, because of the absence of new jobs; particularly hard hit were the young members of discriminated-against racial minorities.

These high levels of unemployment accentuate the problems of the cycle of poverty. Incomes are low, people are dependent on state benefits, public services, and public housing, and the stresses and strains of coping multiply – in a society which tends to consider unemployment as a result of individual failings rather than structural factors. Health problems increase, and there are many stimuli to crime, petty or otherwise. Most importantly, perhaps, long-term unemployment and deprivation encourage the development of a sense of alienation from society – a society which, because of the spatial segregation, is 'out there'. The representatives of that society are not trusted – especially the police, and the officers of local government and welfare services. Distrust leads to conflict. At the individual level, this can be contained by the society, but occasionally it may erupt into group conflict. The when and where of such conflict cannot be predicted.

The problems of the inner city provide the context, but cultural variations mean that it is more threatening among some groups and in some places than others. But some trigger event is needed to initiate major eruptions, such as the inner city 'race riots' in Britain in summer 1981, most of which were stimulated by the relationships between groups of youths and the police. As with most conflicts within the city, these are local. They may spread to some extent but they are invariably ad hoc and unorganized. They represent reactions to conditions rather than well-thought-out movements for major social and economic reforms.

Countering the cycle: Urban programmes

How can the problems of the cycle of poverty be corrected? Where in the cycle should attempts be made to eliminate deprivation? Several policies have been proposed and enacted, most of them aimed at the populations of areas rather than at the individuals *per se*. Few have even recognized the fundamental problem that deprivation results from position in the class system.

A widely held assumption associates much deprivation with environmental causes, from which it follows that removal of the environmental problems would lead to elimination of the cycle of poverty and its symptoms, such as crime. Much of the rationale for public housing programmes rests on this assumption, and the slum-clearance policies in many countries were introduced to achieve social improvement through environmental change: nice places make nice people!

The public housing programmes were partly successful. In terms of health, the new living environments were much superior to those replaced and many physical ailments were alleviated, especially in children. But unfortunate consequences, some of which could have been foreseen, were introduced too. In many cases, for example, the rents for the new homes were much higher than those for the old, thereby accentuating the problems of poverty. (This meant that a new round of policies was needed which either subsidized rents, as in United States public housing, or provided rent supplements to poor families, as in Britain.) And where the housing was built in the suburbs, the cost of the journey to work was increased too, sometimes considerably, with the same effects on real incomes. The later decentralization of employment opportunities has meant that to find and keep jobs many individuals living the cycle of poverty have had to invest in their own motor vehicles.

The emphasis on housing improvement as the solution to deprivation meant that many of the new suburban housing estates were provided with no other facilities. In parts of suburban Glasgow, for example, large public housing estates lacked both shops and schools for many years after their completion, which meant expensive journeys, at least to the former. Community facilities were also lacking, including public houses, which again provided an external focus for social life and encouraged the pursuit of leisure on the streets by teenagers. The lack of comprehensive planning for a community life was the main problem, plus of course the difficulty that planners experience in convincing private firms to locate on the estates. The few shops that did were able to charge monopoly prices, thereby accentuating the problems of living on low incomes.

Much public housing in recent years has involved the redevelopment of inner-city sites, using high-rise apartment blocks. Some of these were widely praised by architectural critics, and innovations included attempts to recreate the street communities of the older dwelling arrangements. But the acclaim was not echoed by the residents, who soon identified many problems of life in the blocks. Some refused to live in them, and a number of blocks rapidly became derelict; indeed some have been demolished, several decades before the loans with which they were built have been repaid. Some of the faults were architectural: research has suggested, for example, that various aspects of the designs created unobserved, unpoliced areas (such as stairwells) which were havens for the idle and provided ideal locales for the commission of crime, much of it petty in itself but irritating in its totality. Other problems reflected unforeseen changes. Some blocks, for instance, had only under-floor central heating systems installed and when these became too expensive for the low-income tenants they switched them off; their alternatives, such as free-standing paraffin heaters, created condensation and other problems which stimulated rapid environmental deterioration. But the majority of the problems were managerial. Families with small children were allocated apartments on upper floors. The balconies proved to be unsatisfactory as play areas and mothers were not prepared to have their children playing outside, many floors below, where they were not easily observed and supervised. Many young children were kept indoors all day, therefore, with consequences for the mental and physical health of both themselves and their parents. And finally, unlike French apartment blocks with their *concierges*, and high-class developments in most countries with doormen and restricted access to the buildings, these public housing blocks lacked supervision. This allowed them to become havens for vandals, with consequences for lift maintenance, for example. Rather than solve the managerial problems, however, most housing authorities have decided not to build any more high-rise blocks but to return to high-density, low-rise developments.

Improving the housing situation by providing new homes alleviated some of the problems of the cycle of poverty, therefore, but it ignored others and even exacerbated some, such as the poverty itself. And so other points in the cycle became the foci for attention, such as education. People in the cycle of poverty were under-achieving according to the ethos of capitalist society, and so, the argument went, they should be provided with a better education to compensate for the disabilities of their home backgrounds. Thus in Britain, for example, inner-city school catchment areas were designated as Educational Priority Areas and the schools received 'positive discrimination' in the form of teaching resources, income supplements aimed at attracting better teachers, and preferential provision of nursery education facilities. It is doubtful whether the relatively small input of extra resources has had much impact: one of the problems has been that in focusing on providing resources for areas, many of the deprived do not benefit while many of those not suffering particularly from the cycle of poverty do.

Crime, it has been argued here, is a symptom of the cycle of poverty and it is unlikely that it can be removed independently of any action which removes its basic causes. Greater police activity in the known areas of crime commission can effect some improvement, but not with the crimes committed in the home as consequences of the stresses of life in poverty. (The use of social workers similarly is a palliative only.) Like other programmes, this is ameliorative problem-solving: it identifies a problem but tackles the surface manifestation rather than the root cause.

Housing, education, crime prevention and other 'positive discrimination' programmes aimed at aiding those living in the cycle of poverty were independent reactions to the perceived problems of deprivation. Their relative failures led to more comprehensive approaches such as the Model Cities projects in the United States and the so-called Urban Programme in Britain. These aimed at community revitalization, at upgrading the existing inner-city environments through grants for housing and other improvements, and were intended to involve members of the community working for their own betterment. In addition, in many places these programmes were also aimed at defusing growing problems of race relations: Britain's Urban Programme, for example, was announced by the Prime Minister as a reaction to Enoch Powell's 'rivers of blood' speech. But the financial input has been relatively small and the grants for housing and environmental improvements have been denied the worst areas, where the properties have only short life expectancies but where many of the most deprived groups, including racial minorities, live. The community workers employed to stimulate the local involvement were in several cases led by their experiences to produce a reappraisal of the situation which saw the cycle of poverty as part of the structural problems of capitalist society. They concluded that only a reformulation of society could lead to removal of poverty and its manifold expressions.

Since the late 1970s, several countries have introduced policies aimed at attracting jobs back to the increasingly derelict inner cities – their goal is countering counter-urbanization. Many of these are based on the concept of the enterprise zone, introduced by Peter Hall but itself a development of the free trade zones and tax-free enclaves already operated in several countries. Conventional wisdom at the time was that most new jobs are being created not by large corporations but by small firms, so policies were needed to stimulate the latter and to encourage the creation of many more of them (i.e. the seed-bed growth component in Figure 4.3). Planning regulations and the weight of tax and labour laws are hindering the needed enterprise behind new initiatives, it was argued; to release the latent entrepreneurial talents, enterprise zones offer many fewer restrictions, and thereby seek to stimulate new factory and job formation. Whether they will do is doubtful. The incentives reduce costs slightly, but almost inconsequentially compared to the major labour cost differentials between, say, western Europe and Taiwan. Without other controls (e.g. on imports and exchange rates) it is unlikely that enterprise zones will stimulate many new jobs, in countries with many millions of unemployed, in unattractive inner city areas.

The cycle of poverty is not an independent element within urban society. It has a counterpart: the cycle of affluence (Figure 9.2). Mobility between the two is possible (usually through intermediate cycles which are not identified in the diagram), but a major net migration from poverty to affluence is only likely in a period of rapid economic growth for the society as a whole. The members of the cycle of affluence wish to protect their positions, which is why they are in conflict with those in the cycle of poverty and why they fail to grant the latter the real means of escaping poverty – higher real incomes and job security. The concessions that they may feel politically impelled to make – as with housing and educational priority areas – can be clawed back with other policies, such as higher ('economic') rents. And relative deprivation can always be maintained by insisting on parallel improvements in both cycles: with education, for example, better facilities in the cycle of poverty schools can be offset by better facilities in those of the cycle of affluence too, so that students in the latter maintain their educational superiority and obtain the qualifications which provide the entrée to the well-paid, relatively secure jobs. Late capitalist society, as Hirsch points out, is a positional society: if everybody in a crowd has the resources to afford one brick to stand on, then those with wealth and power will ensure that they can afford two.

The solutions to the cycle of poverty do not lie in programmes aimed

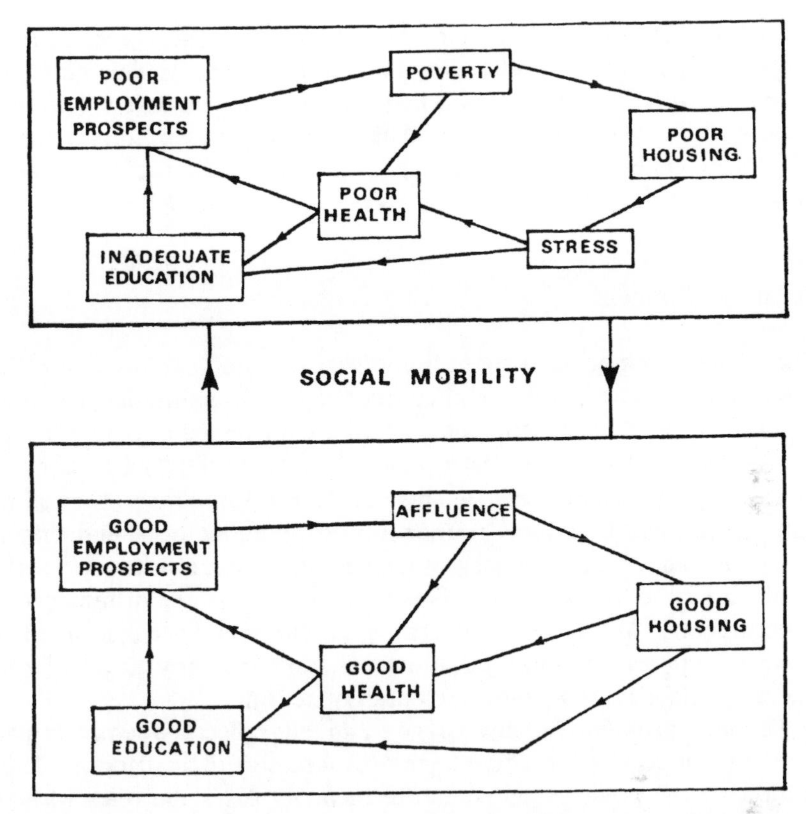

Figure 9.2 The cycle of poverty and its links with the cycle of affluence.

at improving conditions in areas of multiple deprivation, therefore, desirable though those improvements may be as a part of a general policy of environmental upgrading. The only lasting answers lie in changing the distribution of income and access to employment opportunities, which involve major restructuring of society and not just cosmetic surgery to its external manifestations. Thus the sorts of programmes discussed here cannot remove the basic inequalities of capitalist urban societies. This creates a major ethical problem for many researchers. There clearly are areas of major deprivation and there are many things which can be done to improve the quality of life for those forced by the housing system to live in those deprived conditions; such improvements are becoming more desperately needed as employment opportunities are withdrawn from the inner cities. But by concentrating on such improvements, the continuation of the cycle of poverty, and its counterpart in the cycle of affluence, is ensured. Increasingly, urban planners and associated social scientists are deciding that removing the problems perceived in the cycle of poverty involves changing the nature of society. This puts them in conflict with their employers, who only want changes that will preserve society as it currently operates. The conflict between classes that is endemic to capitalism, and probably to state capitalism too, is now being fought within the state edifice.

Urban government

The state is a major participant in modern societies, capitalist or otherwise. In most countries the detailed operation and maintenance of cities is devolved to local governments, which are presumed to perform three basic functions in society: they protect individual liberty by offering a mosaic of counterbalances to a single centralized autocracy; they stimulate involvement by a large proportion of the population; and they increase efficiency because local governments are better able to tap peculiar local variations in needs for certain services than are distant central governments. Most local governments are closely constrained by centrally imposed regulations, however, and in many areas of their operation have relatively little freedom of action.

The main problems facing local governments reflect the demands made on them for services, relative to the resources at their disposal. Those subject to the greatest demands are in many cases the ones with the smallest resources. This is particularly the case in the United States and other countries where the inner-city areas, containing most of the worst elements of the cycle of poverty, cannot draw on the resources of the suburban population in the cycle of affluence, who have placed themselves the other side of municipal boundaries and make relatively small contributions to city budgets. Municipal balkanization condemns the poor to finance their own needs, as indeed it would do if the rich were in the inner city and the poor were in the suburbs.

The solution to this problem would seem to be a system of metropolitan government, with the entire urban area controlled by a single authority raising its finance from the whole population. This solution has been attempted in Britain, with the creation of the Greater London Council in the early 1960s and the reorganization of local government elsewhere in the country in the 1970s. In general, however, the functional urban areas have been underbounded, allowing some commuters to flee beyond the municipal frontier into lower-rated rural counties where the demands for services are relatively low. Although this is now a minor problem compared to the earlier situation, it highlights the difficulties of defining urban areas for administrative purposes and to some commentators indicates the need for even broader regional governments which combine towns and cities with their rural hinterlands. Even if the latter type of reorganization were introduced, however, there would still be differences between and within regions in their resources and their abilities to service the demands made on them. As pointed out in Chapter 5, most countries have towns which are in economic decline and have a higher share of the country's economic and social problems than they have of its population; these are less able to finance schemes than are the growing cities whose economies are booming and whose workforce is nearly all in employment.

Britain's centralized authority includes a national government which has been prepared on several occasions to impose a reorganization of local government, even if this displeases the existing power structure in the cities and counties. Governments in other countries have not been prepared to take such potentially unpopular political notions, however: in the USA, for example, suburbs are major centres of power within State congresses and it is not in the interests of those running the latter to alienate their constituents. Thus in most States metropolitan government is only introduced if the residents want it; referenda indicate that in general they do not, and where it has been introduced the constraints to action have been tight.

Many local governments are facing a chronic financial crisis. This reflects on their available resources. Taxes on property values have for long been a major source of revenue for local governments. These are regressive, in that they hit hardest at the relatively poor because there is no minimum value below which the taxes are not paid and the relationship between income and property value is non-linear. (The former problem has to some extent been overcome in Britain by the system of rate rebates.) Thus many inner-city governments have a very shallow tax base on which to draw, and their problems are increasing as the industrial and commercial enterprises which formerly provided much of their income decentralize to the suburbs and their cheaper taxes.

Rather than alter the source of revenue for local governments and create a more progressive local tax system, many national governments have sought to shore up the local authorities with various grant systems. These are usually so constructed that the national governments are given even closer control over the operations of the recipients. To the extent that they are obtained from the progressive taxes which are the main sources of revenue to national governments and to the extent that they are allocated according to needs, then these grant systems redistribute money from the affluent to the poor. But the extent of the redistribution is generally slight and the size of the grants does not allow the local governments to meet all of their demands. There is little evidence, therefore, that local governments are solving their local problems, let alone those that impinge on a mosaic of adjacent local governments.

Conclusions

The city is the focus of society's social and economic problems. Within it, many of the problems are spatially concentrated. This is because they reflect the divisions of society, divisions which are manipulated and maintained by a series of mechanisms that allocate people to their place in both the social and the spatial structure. The manipulation is undertaken by those in control of the society, who are determined to maintain its morphology, to protect the social and economic systems and their own positions within them.

Some of the problems, such as those of pollution and traffic management, will be tackled because the powerful stand to benefit (perhaps more than the weak) and because at least part of the costs can be externalized. But there is greater reluctance to tackle some of the others, because to solve them would be to create a potential threat to the stability of the society. Thus although some improvements have been made to the absolute conditions experienced by those living in the cycle of poverty, for example, little has been done to improve the competitive position of these people relative to those in the cycle of affluence. Capitalist societies are built on inequalities which are reflected in the spatial arrangement of cities. Changing the spatial arrangement does not remove the inequalities, however; it simply rewrites them. Urban planning is not an agent of major social and economic change, therefore, but rather one of preservation. Some reorganization does occur, but the fundamental structure of society remains.

Further Reading

The selection of material presented here is only a small, non-random sample of the very large literature on which this book draws. In particular, there is little reference to individual pieces of research.

Chapter 1

For a general history of human geography, see R. J. Johnston, *Geography and Geographers: Anglo-American Human Geography since 1945* (Edward Arnold, London, 1979); for urban geography see Chapter 1 of B. J. L. Berry and J. D. Kasarda, *Contemporary Urban Ecology* (Macmillan, New York, 1977) and Chapter 1 of D. T. Herbert and R. J. Johnston, *Geography and the Urban Environment*, Vol 1. (John Wiley, London, 1978). The reference to Hartshorne is to R. Hartshorne, *Perspective on the Nature of Geography* (Rand McNally, Chicago, 1959).

There are many textbooks on urban geography. For a recent example of the 'traditional' approach see J. H. Johnson, *Urban Geography* (Pergamon Press, London, second edition, 1972), whilst a more specialized treatment of historical urban geography can be found in J. E. Vance, Jr, *This Scene of Man* (Harper & Row, New York, 1977). A widely used text which spans 'traditional' and 'modern' approaches is H. Carter, *The Study of Urban Geography* (Edward Arnold, London, second edition, 1975); the 'modern' methods are typified in M. H. Yeates and B. J. Garner, *The North American City* (Harper & Row, New York, second edition, 1976) and a book of readings edited by B. J. L. Berry and F. E. Horton, *Geographic Perspectives on Urban Systems* (Prentice-Hall, Englewood Cliffs, New Jersey, 1970). The behavioural perspective is exemplified in D. T. Herbert, *Urban Geography: A Social Approach* (David & Charles, Newton Abbott, 1972) and L. J. King and R. G. Golledge, *Cities, Space and Behavior: The Elements of Urban Geography* (Prentice-Hall, Englewood Cliffs, New Jersey, 1978). The general thesis presenting geography as a spatial science is well illustrated by R. Abler, J. S. Adams and P. R. Gould, *Spatial Organization* (Prentice-Hall, Englewood Cliffs, New Jersey, 1971) and

P. Haggett, A. D. Cliff and A. E. Frey, *Locational Analysis in Human Geography* (Edward Arnold, London, 1977). The major critiques of the spatial science approach to urban geography were led by D. Harvey, *Social Justice and the City* (Edward Arnold, London, 1973); see also D. Gregory, *Ideology, Science and Human Geography* (Hutchinson, London, 1978). Other important books in this critique include M. Castells, *The Urban Question* (Edward Arnold, London, 1977), K. R. Cox, *Urbanization and Conflict in Market Societies* (Maaroufa Press, Chicago, 1977), M. Harloe, *Captive Cities* (John Wiley, London, 1977), R. E. Pahl, *Patterns of Urban Life* (Longman, London, 1970) and *Whose City?* (Penguin, Harmondsworth, second edition, 1975), and L. K. Sawer and W. Tabbs, *Marxism and the Metropolis* (Oxford University Press, New York, 1977).

Chapter 3

A seminal work by a geographer on urban origins is P. Wheatley, *Pivot of the Four Quarters* (Aldine Press, Chicago, 1971); see also the paper by H. Carter in *Progress in Human Geography*, **1**, 1977, 12–32. The classic presentation of central place theory is W. Christaller, *Central Places in Southern Germany* (Prentice-Hall, Englewood Cliffs, New Jersey, 1966); a good exposition is C. R. Lewis, 'Central place analysis', in Open University, *Fundamentals of Human Geography, Section II: Spatial Analysis, Point Patterns* (The Open University, Milton Keynes, 1977); see also B. J. L. Berry, *The Geography of Market Centers and Retail Distribution* (Prentice-Hall, Englewood Cliffs, New Jersey, 1967). On urban systems under mercantilism the best single source is J. E. Vance, *The Merchant's World: The Geography of Wholesaling* (Prentice-Hall, Englewood Cliffs, New Jersey, 1970); important papers include A. J. Rose, 'Dissent from down-under', *Pacific Viewpoint*, **7**, 1966, 1–27, and A. F. Burghardt, 'A hypothesis about gateway cities', *Annals of the Association of American Geographers*, **61**, 1971, 269–85. On periodic markets, see the recent review by R. J. Bromley in D. T. Herbert and R. J. Johnston, *Geography and the Urban Environment*, Vol. 3 (John Wiley, London, 1980); for reviews of urban systems, see the essays in R. Jones, *Essays on World Urbanization* (George Philip, London, 1975), and K. Dziewonski, 'Urbanization and settlement systems', *Geographia Polonica*, **38**, 1978. On cities and development, see B. Roberts, *Cities of Peasants* (Edward Arnold, London, 1978).

Chapter 4

For an introduction to the literature on industrial location, see D. M. Smith, *Industrial Location* (John Wiley, New York, 1971); for the urban multiplier, see the work summarized in A. Pred, *City-Systems in Advanced Economies* (Hutchinson, London, 1977). Of the many city classifications the most cited is probably C. A. Moser and W. Scott, *British Towns* (Oliver & Boyd, Edinburgh, 1962); for an encyclopedic presentation, see B. J. L. Berry, *City Classification Handbook* (John Wiley, New York, 1972). Innovation in the urban system is treated in B. T. Robson, *Urban Growth: An Approach* (Methuen, London, 1973), and the tertiary and quaternary sectors are outlined in P. W. Daniels, *Office Location: An Urban and Regional Study* (Bell, London, 1975). A good collection of essays on city systems is L. S. Bourne and J. W. Simmons, *Systems of Cities* (Oxford University Press, New York, 1978).

Chapter 5

On urban and regional planning, see P. Hall, *Urban and Regional Planning* (Penguin, Harmondsworth, 1974); the classification of planning styles is taken from B. J. L. Berry, *The Human Consequences of Urbanization* (Macmillan, London, 1973). Over-urbanization can be studied through the essays in G. W. Breese, *The City in Newly Developing Countries* (Prentice-Hall, Englewood Cliffs, New Jersey, 1969) and in T. G. McGee, *The Urbanization Process in the Third World* (Bell, London, 1971); a recent collection of relevant essays is B. J. L. Berry, *Urbanization and Counter-Urbanization* (Sage Publications, Beverly Hills, 1977). The problems of urban size are dealt with in H. W. Richardson, *The Economics of Urban Size* (Saxon House, Farnborough, 1973); see also the arguments of A. G. Gilbert, 'The arguments for very large cities reconsidered', *Urban Studies*, **13**, 1976, 23–34, R. J. Johnston, 'Observations on accounting procedures and urban-size policies', *Environment and Planning A*, **8**, 1976, 327–40, and I. Hoch, 'Income and city size', *Urban Studies*, **9**, 1972, 299–328. A general treatment of urban-size policies is L. S. Bourne, *Regulating Urban Systems* (Oxford University Press, London, 1975); see also H. Stretton, *Urban Planning* (Oxford University Press, London, 1978); and for the problems of particular cities, see P. Hall, *The World Cities* (Weidenfeld & Nicolson, London, second edition, 1977). Problem regions are dealt with in A. J. Brown and E. M. Burrows, *Regional Economic Problems* (George Allen

& Unwin, London, 1977). On the prisoner's dilemma, see M. Taylor, *Anarchy and Cooperation* (John Wiley, London, 1976).

Chapter 6

K. R. Cox, *Conflict, Power and Politics in the City* (McGraw-Hill, New York, 1973) presents an introduction to externalities; for classes in the city, see M. Castells, *City, Class and Power* (Macmillan, London, 1978). Immigration and cities is treated in D. Ward, *Cities and Immigrants* (Oxford University Press, New York, 1971). The topic of neighbourhood and socialization is treated by D. Timms, *The Urban Mosaic* (Cambridge University Press, London, 1971) and the chapters by D. Timms and by D. T. Herbert in D. T. Herbert and R. J. Johnston, *Urban Social Areas* (2 Vols. John Wiley, London, 1976). Social contacts and communities can be studied in two books by G. D. Suttles, *The Social Order of the Slum* (University of Chicago Press, Chicago, 1969) and *The Social Construction of Communities* (University of Chicago Press, Chicago, 1972) and in H. J. Gans, *People and Plans* (Basic Books, New York, 1969); a good case study is D. Ley, *The Inner-City Ghetto as Frontier Outpost* (Association of American Geographers, Washington, 1974). On chain migrations, see the essay by A. D. Trlin in R. J. Johnston, *Urbanization in New Zealand* (Reed, Wellington, 1973); the essay by F. W. Boal in D. T. Herbert and R. J. Johnston, *Social Areas in Cities* (John Wiley, London, 1976) discusses ghettos in general, and a particular case is discussed in H. J. Gans, *The Urban Villagers* (The Free Press, New York, 1962). On territoriality, see R. Ardrey, *The Territorial Imperative* (Atheneum Press, New York, 1966). The material on apartheid city is taken from R. J. Davies's paper in P. Adams and F. K. Helleiner, *International Geography*, Vol. 2 (University of Toronto Press, Toronto, 1972); for American exclusionary planning, see M. N. Danielson, *The Politics of Exclusion* (Columbia University Press, New York, 1976).

Chapter 7

Two good compendia of essays on the patterns within cities are L. S. Bourne, *The Internal Structure of Cities* (Oxford University Press, New York, 1971) and K. P. Schwirian, *Comparative Urban Structure* (D. C. Heath, Lexington, Massachusetts, 1974). Factorial ecologies are dealt with in the essay by R. J. Johnston in D. T. Herbert and R. J. Johnston, *Social Areas in Cities* (John Wiley, New York, 1976); see also the essay by Bourne in that book. For housing market operations, see essays by D. Harvey in H. M. Rose and G. Gappert, *The Social Economy of Cities* (Sage Publications, Beverly Hills, 1975) and in R. Peel, M. Chisholm and P. Haggett, *Processes in Physical and Human Geography* (Heinemann, London, 1975).

For squatter settlements, see P. Lloyd, *Slums of Hope?* (Penguin, Harmondsworth, 1979) and D. J. Dwyer, *People and Housing in Third World Cities* (Longman, London, 1975). The concept of housing classes was introduced in J. Rex and R. Moore, *Race, Community and Conflict* (Oxford University Press, London, 1967); see also R. E. Pahl, *Whose City?* (Penguin, Harmondsworth, second edition, 1975) and J. E. Vance in D. T. Herbert and R. J. Johnston, *Social Areas in Cities* (John Wiley, London, 1976). On particular institutions, see the papers by M. J. Boddy and P. R. Williams in *Transactions of the Institute of British Geographers*, NSI, 1976, for building societies, by R. Palm in D. T. Herbert and R. J. Johnston, *Geography and the Urban Environment*, Vol. 2 (John Wiley, London, 1979) for estate agents, and by F. Gray in *Transactions of the Institute of British Geographers*, NSI, 1976, for public housing allocation; urban renewal is treated in J. Lambert, C. Paris and B. Blackaby, *Housing Policy and the State* (Macmillan, London, 1978). On intra-urban migration, see the review by J. S. Adams and K. Gilder in D. T. Herbert and R. J. Johnston, *Social Areas in Cities* (John Wiley, London, 1976); a chapter by R. A. Murdie in the same book treats the spatial aspects of urban residential patterns, whereas the basis for the economic arguments is W. Alonso, *Location and Land Use* (Harvard University Press, Cambridge, 1965); the geography of land values and land-value changes is discussed by R. J. Johnston in D. A. Lanegran and R. Palm, *Invitation to Geography* (McGraw-Hill, New York, second edition, 1978). The general evolutionary model derives from papers in L. F. Schnore, *The Urban Scene* (The Free Press, New York, 1965) and R. J. Johnston in C. Board *et al.*, *Progress in Geography*, Vol. 4 (Edward Arnold, London, 1972).

Chapter 8

On the evolution of urban industries, see the two case studies in P. Hall, *The Industries of London since 1861* (Hutchinson, London, 1962) and J. E. Martin, *Greater London, An Industrial Geography* (Bell, London, 1966); recent changes are analysed in the paper by D. B. Massey and R. A. Meegan, *Urban Studies*, **15**, 1978, 273–88. The office sector is treated in several books by J. Goddard, including *Office Linkages and Location* (Pergamon Press, Oxford, 1973); see also P. W. Daniels, *Spatial Patterns* *of Office Growth and Location* (John Wiley, London, 1979). There is a very large literature on urban retailing: a recent review is by J. A. Dawson and D. Kirby in D. T. Herbert and R. J. Johnston, *Geography and the Urban Environment*, Vol. 3 (John Wiley, London, 1980). On public facilities, see B. H. Massam, *Location and Space in Social Administration* (Edward Arnold, London, 1975); the political issues are discussed in R. J. Johnston, *Political, Electoral and Spatial Systems* (Oxford University Press, London, 1979).

Chapter 9

A general book on urban problems is R. Davies and P. Hall, *Issues in Urban Society* (Penguin, Harmondsworth, 1978). On urban environments, see the two collections of essays in T. R. Detwyler and M. G. Marcus, *Urbanization and Environment* (Duxbury Press, Belmont, California, 1972) and B. J. L. Berry and F. E. Horton, *Urban Environmental Management* (Prentice-Hall, Englewood Cliffs, New Jersey, 1974). On the role of planning in society and the conflicts involved, see T. A. Broadbent, *Planning and Profit in the Urban Economy* (Methuen, London, 1977) and J. M. Simmie, *Citizens in Conflict* (Hutchinson, London, 1974). On the neighbourhood effect, see the examples regarding crime in J. Baldwin and A. E. Bottoms, *The Urban Criminal* (Tavistock, London, 1976), regarding education in M. Rutter, *Fifteen-Thousand Hours* (Open Books, London, 1979), and regarding political attitudes in P. J. Taylor and R. J. Johnston, *Geography of Elections* (Penguin, Harmondsworth, 1979). On the inner-city problem, see the essay by N. J. Thrift in D. T. Herbert and R. J. Johnston, *Geography and the Urban Environment*, Vol. 2 (John Wiley, London, 1979). Architectural problems are discussed in O. Newman, *Defensible Space* (Architectural Press, London, 1973) and C. Mercer, *Living in Cities* (Penguin, Harmondsworth, 1975). The urban programme in Britain is discussed in J. Edwards and R. Batley, *The Politics of Positive Discrimination* (Tavistock, London, 1978), and urban governmental problems are outlined in R. J. Johnston, *Political, Electoral and Spatial Systems* (Oxford University Press, London, 1979). The concept of the positional society is introduced in F. Hirsch, *Social Limits to Growth* (Routledge & Kegan Paul, London, 1977); the review of the balanced neighbourhood concept is by W. Sarkissian in *Urban Studies*, **13**, 1976, 231–46.

Since the first edition of this book was written in 1979, much relevant material has appeared.

The general format of this book has been used to develop a case study of city and society in one country – the United States: R. J. Johnston, *The American Urban System* (St. Martin's Press, New York and Longman, London, 1982). A general study of the UK is provided in R. J. Johnston and J. C. Doornkamp, *The Changing Geography of the United Kingdom* (Methuen, London, 1983). The economic theory around which this book is built is set out fully and clearly in D. Harvey, *The Limits to Capital* (Blackwell, Oxford, 1982). The most recent general texts are J. R. Short, *An Introduction to Urban Geography* (Routledge and Kegan Paul, London, 1984), and B. Badcock, *Unfairly Structured Cities* (Blackwell, Oxford, 1984).

For recent work on urban systems see: S. D. Brunn and J. F. Williams, *Cities of the World* (Harper and Row, New York, 1983); L. S. Bourne, K. Dziewonski and R. Sinclair, *Urbanization and Settlement Systems* (Oxford University Press, New York, 1984); and B. J. L. Berry, *Contemporary Urbanization* (St. Martin's Press, New York, 1983).

On residential patterns in cities, recent books include: K. Bassett and J. R. Short, *Housing and Residential Structure* (Routledge and Kegan Paul, London, 1980); P. L. Knox, *Urban Social Geography* (Longman, London, 1982); D. Ley, *A Social Geography of the City* (Harper and Row, New York, 1983); and P. E. White, *The West European City: A Social Geography* (Longman, London, 1984). On political aspects see: K. R. Cox and R. J. Johnston, *Conflict, Politics and the Urban Scene* (Longman, London, 1982); R. J. Johnston, *Residential Segregation, the State and Constitutional Conflict in American Suburbia* (Academic Press, London, 1984); and J. R. Short, *The Urban Arena* (Macmillan, London, 1984). Urban historical geography is treated in J. W. R. Whitehand, *The Urban Landscape* (Academic Press, London, 1981) and H. Carter, *An Introduction to Urban Historical Geography* (Edward Arnold, London, 1983), and urban planning in A. Sutcliffe, *Towards the Planned City* (Blackwell, Oxford, 1981). For the issues facing urban government see K. Newton, *Balancing the Books* (Sage, London, 1981), and on the physical environment I. Douglas, *The Urban Environment* (Edward Arnold, London, 1983).

On the problems of the inner city see: P. Hall, *The Inner City in Context* (Heinemann, London, 1982) and P. Harrison, *Inside the Inner City* (Penguin, London, 1983).

For the urban geography of socialist societies see: I. Szelenyi, *Urban*

Inequalities under State Socialism (Oxford University Press, Oxford, 1983) and the essays in R. A. French and F. E. I. Hamilton, *The Socialist City* (John Wiley, Chichester, 1979).

Index

agglomeration economies, 92–93
apartheid, 175–9

balanced neighbourhoods, 274–6
basic industries, 103–105
behaviourist critique, 23
bureaucracy, 49–50, 149

capitalism, 34–42, 42–8
central place theory, 17–19 (*see also* hierarchies)
centralization, 39, 45, 111, 115–19
city problems, 260–64
city sizes, 16–19, 132–41
class, 42–6, 153–62, 201–208
colonialism, 53–7, 67–75
communities, 262–3, 270–76
commuting, 239–42, 245
concentration, 38–9, 45, 112–15
counter-urbanization, 143–6
culture, 227–8
cycle of poverty, 273, 276–84

deconcentration, 117–19
diseconomies of scale, 134
distancing, 163–83

education, 158–9, 165–7, 182–3, 271–3
environmental problems, 260–61, 264–6
exclusionary zoning, 179–83
external economies, 91–93
externalities, 164–66

factorial ecology, 21, 188–9
factory system, 86–99

gateway cities, 71–5, 80–85, 108
gradients, 219–24

health care, 256–7
hierarchies, 53–7, 59–63, 68–70, 76–9, 245–55
housing classes, 188, 201–203
housing markets, 189–97, 203–214

immigration, 159–61, 171–5
industrial capitalism, 35–8, 83–109
industrial location, 19, 93–6, 235–9
inner city, 278–9
innovation, 105–7, 139
interaction, 167–70
internal economies, 89–91
internal structure of cities, 20–22

late capitalism, 38–42, 109–21
lifeworld, 230–31
logical positivism, 14

market circuits, 58–61
marriage, 169
mercantilism, 31–4, 43, 47, 57–75, 98–101
migration, 65–6, 212–14
monopoly, 38–9
mortgages, 203–8

neighbourhood effect, 166–7, 271–6
new towns, 96–8, 108, 142
nonbasic industries, 103–105
nonresidential land uses, 20, 233–59

offices, 242–5
oligopoly, 38–9
overurbanization, 126–31

periodic marketing, 58–61, 79
planning, 14, 124–6, 137–43, 250–55, 266–70
pre-industrial urbanization, 85–7
prices, 250–55
primate city, 16–17, 80–85
prisoner's dilemma, 136–7
problem towns, 146–8
property values, 170–71, 189–91
public housing, 197–8, 208–12, 279–81
public services, 257–8

quaternary sector, 41, 119–21

rank-distribution societies, 29, 42–43, 46, 53–7, 157
rank-size rule, 16, 77, 99–100
reciprocal societies, 28–9
residential segregation, 153–84
retailing, 114, 245–55

scale economies, 89–93, 132
sectors, 215–19
seed-bed process, 106–7, 237, 281
segregation, 175–83, 185–89, 203–214
socialism, 49–50, 149–50
socialization, 166–7
spatial organization, 14–25, 261–3 266–70
spatial patterns, 215–28
sprawl, 142–44
squatter settlements, 196–7
state, 40–41, 46–8, 120, 137–41, 197–201, 257–8
state capitalism, 48–9, 147–50, 210–12
structuralist critique, 24–5
subsistence societies, 28–9
suburbanization, 180–83, 226–27, 239–40

territoriality, 175–9
tertiary sector, 41, 119–21
townscape, 228–30

urban geography, 13–27
urban government, 179–83, 263, 284–6
urban growth, 101–107
urbanization, 65–6, 126–31
urban programmes, 279–84
urban systems, 15–20, 57–66, 96–101, 107–9, 112–22, 147–50

zones, 215–24